Persistent, Bioaccumulative, and Toxic (PBT) Chemicals

Technical Aspects, Policies, and Practices

Adam D. K. Abelkop

John D. Graham

Todd V. Royer

CRC Press
Taylor & Francis Group
Boca Raton London New York

CRC Press is an imprint of the
Taylor & Francis Group, an **informa** business

CRC Press
Taylor & Francis Group
6000 Broken Sound Parkway NW, Suite 300
Boca Raton, FL 33487-2742

First issued in paperback 2017

ISBN-13: 978-1-4822-9877-2 (hbk)
ISBN-13: 978-1-138-79294-4 (pbk)

Library of Congress Cataloging-in-Publication Data

Abelkop, Adam D. K.
    Persistent, bioaccumulative, and toxic (PBT) chemicals : technical aspects, policies, and practices / Adam D.K. Abelkop, John D. Graham, Todd V. Royer.
        pages cm
    "A CRC title."
    Includes bibliographical references and index.
    ISBN 978-1-4822-9877-2 (hardcover : alk. paper) 1. Persistent pollutants. 2. Persistent pollutants--Bioaccumulation. 3. Persistent pollutants--Measurement. 4. Poisons--Measurement. 5. Bioaccumulation. I. Graham, John D. II. Royer, Todd V. III. Title.

TD196.C45A34 2015
628.5'2--dc23   2015034553

Visit the Taylor & Francis Web site at
http://www.taylorandfrancis.com

and the CRC Press Web site at
http://www.crcpress.com

Contents

List of abbreviations

AA	alternatives analysis
AOP	Adverse Outcome Pathway
BAF	bioaccumulation factor
BCF	bioconcentration factor
BMF	biomagnification factor
CAP	Chemical Action Plan
CAS	Chemical Abstracts Service
CDC	Centers for Disease Control and Prevention
CEPA	Canadian Environmental Protection Act
CLRTAP	Convention on Long-Range Transboundary Air Pollution
CMP	Chemicals Management Plan
CMR	carcinogenic, mutagenic, or toxic to reproduction
CSCL	Chemical Substances Control Law
CSIA	Chemical Safety Improvement Act
CSR	Chemical Safety Report
DDE	dichlorodiphenyldichloroethylene
DDT	dichlorodiphenyltrichloroethane
DEHP	bis(2-ethylhexyl) phthalate
DEQ	Department of Environmental Quality (Oregon)
DfE	Design for the Environment
DOE	Department of Ecology (Washington)
DSL	Domestic Substances List
DTSC	Department of Toxic Substances Control
DYNAMEC	Dynamic Selection and Prioritisation Mechanism for Hazardous Substances
EC	Environment Canada
EC$_{50}$	Half-Maximal Effective Concentration
ECHA	European Chemicals Agency
ECOSAR	Ecological Structure Activity Relationships
EDC	endocrine-disrupting chemical
EPA	Environmental Protection Agency
EPI Suite	Estimation Program Interface Suite
EPS	expanded polystyrene

EU	European Union
GHS	Globally Harmonized System
GLWQA	Great Lakes Water Quality Agreement
HBCDD	hexabromocyclododecane
HC	Health Canada
HCB	hexachlorobenzene
HCBD	hexachlorobutadiene
HCH	hexachlorocyclohexane
IJC	International Joint Commission
iT	inherent toxicity
K_{oa}	octanol–air partition coefficient
K_{ow}	octanol–water partition coefficient
LC_{50}	median lethal concentration
LD_{50}	median lethal dose
LOEC	lowest observed effect concentration
MCS	multiconstituent substances
MEP	Ministry of Environmental Protection (China)
METI	Ministry of Economy, Trade, and Industry (Japan)
MOA	mode of action
MOE	Ministry of the Environment (Ontario)
NGO	nongovernmental organization
NOEC	no observed effect concentration
NPRI	National Pollutant Release Inventory
OECD	Organization for Economic Cooperation and Development
OPPT	Office of Pollution Prevention and Toxics
OPS	Overarching Policy Strategy
P_{ov}	overall persistence
PACS	Priority Assessment Chemical Substance
PAH	polycyclic aromatic hydrocarbon
PBB	polybrominated biphenyl
PBDE	polybrominated diphenyl ether
PBiT	persistent, bioaccumulative, and inherently toxic
PBT	persistent, bioaccumulative, and toxic
PCB	polychlorinated biphenyl
PCTS	Prohibition of Certain Toxic Substances
PFCA	perfluorocarboxylic acid
PFOA	perfluorooctanoic acid
PFOS	perfluorooctane sulfonate
pFR	polymeric flame retardant
POP	persistent organic pollutant
POPRC	Persistent Organic Pollutants Review Committee
PSL	Priority Substance List
PVC	polyvinyl chloride
QSAR	Quantitative Structure–Activity Relationship

RCRA	Resources Conservation and Recovery Act
REACH	Registration, Evaluation, Authorization, and Restriction of Chemicals
RMO	Risk Management Options
SAICM	Strategic Approach to International Chemicals Management
SAWS	State Administration of Work Safety
SCP	Safer Consumer Products Regulations
SCRAM	Scoring and Ranking Assessment Model
SDS	Safety Data Sheet
SEHSC	Silicones Environmental Health and Safety Council
SGI	Substance Groupings Initiative
SIN	Substitute It Now
SLRA	screening level risk assessment
SNAc	Significant New Activity
SNUR	Significant New Use Rule
SVHC	substance of very high concern
TGD	Technical Guidance Document
TMF	trophic magnification factor
TRI	Toxic Release Inventory
TSCA	Toxic Substances Control Act
TSL	Toxic Substances List
UNECE	United Nations Economic Commission for Europe
UNEP	United Nations Environment Programme
UVCB	unknown or variable compositions, complex reaction products or biological materials
VEL	Virtual Elimination List
vPvB	very persistent, very bioaccumulative
WHO	World Health Organization
WMNP	Waste Minimization National Plan
WMPT	Waste Minimization Priority Tool
WOE	weight of evidence
WSSD	World Summit on Sustainable Development
XPS	extruded polystyrene

Preface

This volume is the culmination of a multiyear, international project examining the science, policies, and practices involved in identifying and governing persistent, bioaccumulative, and toxic chemicals (PBTs). Interest in PBTs has increased in recent years as a result of the implementation of the Registration, Evaluation, Authorization, and Restriction of Chemicals regulation in the European Union and the deliberations about reform of the Toxic Substances Control Act in the United States.

The project originated from a request by the American Chemistry Council to analyze and compare the laws and policies for governing PBTs in Europe, North America, and Asia. Industry, environmental groups, and regulatory agencies have both unique and shared concerns regarding PBTs. Although the project was funded by the American Chemistry Council, we sought and included the views of nonindustry stakeholders, particularly regulatory agencies and nongovernmental organizations involved with PBTs.

Our research involved two primary approaches: the first was extensive interviewing of scientists, regulators, and practitioners from around the world who have relevant experience and expertise. The interviews occurred throughout 2012 and 2013, and a full list of interviewees is included in the Appendix. Interviewees were provided a set of questions addressing the history, current practice, and future of governing PBT chemicals. To promote candor and open discussion, interviewees were informed that their ideas would be conveyed without attribution or quotation. The second approach was to convene a panel of experts on PBT science and policy that included members from Europe and North America. The members of this panel included the authors and contributors to this volume. From the interviews and panel discussions, a report was produced and distributed to a number of experts for peer review. After peer review, the report was finalized and the major findings and recommendations were presented at workshops held in Brussels (December 2013) and Washington, DC (January 2014). Attendees at the workshops expressed interest in a book on the topic of governing PBTs, and we were encouraged to develop the report into a reference book that includes both the science and policy of PBTs. Toward this end, we have added new case studies; provided summary tables of important PBTs; and expanded discussions of various topics, such as the number of PBTs in commerce, weight of evidence approaches, market deselection, and international harmonization.

The information used in this volume is drawn from scientific and policy literature; the interviews of practitioners, regulators, and scientists; and the knowledge and experience of the panel members. It is our hope that this book will serve as a useful reference for those engaged in identifying, studying, or regulating PBTs. It is not an exhaustive academic review, but it is instead designed to serve practitioners and others who require an introduction to the topic, seek an international perspective on PBT policies, or have an interest in the interaction between science and policy development for PBTs.

Acknowledgments

We thank the American Chemistry Council for providing unrestricted financial support for the project. The design and execution of the study, including all findings and recommendations, are fully the responsibility of the authors and contributors. It is important to note that although they are listed as contributors, this book could not have been completed without the efforts of the panel members. We sincerely thank the peer reviewers of the report and the individual experts interviewed during the course of the project for their willingness to share their experience and knowledge. We also thank Ágnes Botos, Kathryn Fledderman, Evan Tyrrell, and Marta Venier for their assistance with the case studies and Yousung Han for his assistance in editing.

Authors

Adam D.K. Abelkop is an associate instructor at the School of Public and Environmental Affairs and a doctoral candidate in Indiana University's Joint PhD in Public Policy program administered by the School of Public and Environmental Affairs and the Department of Political Science. He earned his law degree from the University of Iowa College of Law and completed his undergraduate studies at Wake Forest University. His research focuses on the ways in which scientific and socioeconomic information is incorporated into judicial decision making and risk regulation for environmental and public health protection.

John D. Graham is dean of the School of Public and Environmental Affairs, Indiana University. He earned his undergraduate degree from Wake Forest University, his MA in public policy from Duke University, and his PhD in urban and public affairs from Carnegie-Mellon University. He joined the faculty of the Harvard School of Public Health in 1985 and served as the founding director of the school's Center for Risk Analysis from 1989 to 2001. Dr. Graham's research focuses on risk-based decision making. He is the author or coauthor of more than 10 books and more than 200 articles in academic publications and national journals. In 2001, Dr. Graham was appointed by President George W. Bush as the administrator of the Office of Information and Regulatory Affairs of the US Office of Management and Budget, where he served until 2006.

Todd V. Royer is an associate professor in the School of Public and Environmental Affairs at Indiana University. Dr. Royer holds degrees in ecology from Iowa State University and Idaho State University and was a postdoctoral researcher at the University of Illinois at Urbana-Champaign from 2000 to 2003. His research is aimed at understanding how biogeochemical and ecological factors, including human activities, interact to affect water quality and ecosystem functioning in freshwater systems.

Contributors

Lucas Bergkamp
Hunton & Williams
Brussels, Belgium

Bryan W. Brooks
Department of Environmental
    Science
Baylor University
Waco, Texas

Anna Gergely
Steptoe & Johnson
Brussels, Belgium

George Gray
School of Public Health and Health
    Services
George Washington University
Washington, DC

Gary E. Marchant
College of Law
Arizona State University
Tempe, Arizona

Mallory L. Mueller
School of Public and
    Environmental Affairs
Indiana University
Bloomington, Indiana

Cornelis J. van Leeuwen
KWR Watercycle Research
    Institute
Utrecht, Netherlands

Marco Vighi
Department of Earth and
    Environmental Sciences
University of Milano-Bicocca
Milan, Italy

chapter one

The challenge of identifying, assessing, and regulating PBTs

I. PBTs: An introduction

The term *industrial chemicals* refers to the tens of thousands of substances that are used by businesses on a daily basis throughout the world. As used by regulators, the term *industrial chemicals* generally excludes pharmaceuticals, food additives, biocides, and pest-control products because those chemical applications are covered by unique regulatory processes. Within industry, chemicals are widely used in clothing, furniture, electronics, plastics, paints, cosmetics, and many other consumer products.

A small but important subset of industrial chemicals are of special concern to regulators because they combine three unwanted properties: Once released, they are highly persistent (i.e., they remain in the environment for long periods of time and do not degrade into harmless substances); they have a tendency to bioaccumulate in the tissues of organisms; and they can be toxic at relatively low concentrations when ingested, inhaled, or taken into the body after dermal contact. On a global basis, concerns about such chemicals are causing a wide range of responses by government and industry: more risk assessments to determine whether specific uses are safe for people and the environment; more information (e.g., labeling) for consumers and workers; stringent risk management measures to reduce the frequency and magnitude of releases into the environment; process changes that reduce the need for chemicals; and substitution of different chemicals that do not exhibit these unwanted properties. The responses in different countries are not harmonized, and questions have been raised about whether regulators are responding appropriately to information about the three properties.

How many and which chemicals in widespread industrial and commercial use share these three properties? What measures should be taken to protect public health and the environment from exposures to such chemicals? To what extent are governments around the world identifying these chemicals, and what are they doing about them? What steps can be taken to achieve more international consistency in how such chemicals are regulated? These are the questions that this book addresses.

The term *persistent, bioaccumulative, and toxic* (PBT) is playing an increasingly prominent role in chemical policy (van Wijk et al. 2009). Its emergence as a policy term of art followed the influential term *carcinogen, mutagen, and toxic for reproduction* (CMR) but preceded the relatively new discussion of endocrine disrupting chemicals (EDCs) (Heindel et al. 2013). In the European Union (EU), for example, the Registration, Evaluation, Authorization, and Restriction of Chemicals (REACH) Regulation* treats as "substances of very high concern" (SVHCs) a variety of troublesome chemicals such as CMRs, PBTs,[†] and "substances of equivalent concern" such as EDCs. Chemicals designated SVHCs are subject to priority attention, potential restrictions in specific uses, and eventual phase out. Chemical regulations in Japan, Canada, the United States, and elsewhere also utilize, in various ways, the PBT categorization.

The regulatory history of PBTs suggests that the policy term was first formally used in Japan and was later used in Europe and North America. Publications in the 1960–1970s about various chlorinated substances played a key role in bringing the unwanted properties of persistence and bioaccumulation to the attention of regulators, to nongovernmental organizations, and to industrial managers around the world (van Middelem 1966; Breidenbach et al. 1967; Carson 1962).

Much of this effort has focused on organic compounds. The PBT concept is not easily applied to inorganic substances such as those based on metals or metalloids (e.g., silicon, selenium, and boron) because they are often—by their intrinsic chemical nature—persistent. Nevertheless, metal and metalloid chemicals are increasingly subjected to PBT investigations.

In the process of determining whether a substance is a PBT, separate determinations are first made about a chemical's persistent, bioaccumulative, and toxic properties, and then an overall judgment is made as to whether the substance is a PBT. Although persistence, bioaccumulation, and toxicity are correlated properties (Maeder et al. 2004), there is no single scientific test that reveals whether a particular substance is a PBT. Moreover, the properties of persistence, bioaccumulation, and toxicity are generally presumed to be intrinsic characteristics of a chemical, unaffected by the amount of the substance that has been released into air or water or what the background environmental conditions might be. Although real-world environmental or biomonitoring data (e.g., measuring levels of chemicals in human blood or in Arctic species) are sometimes used to inform PBT determinations, the more commonly used data—at least in regulatory screening exercises—are laboratory tests or

---

* Commission Regulation 1907/2006, Registration, Evaluation, Authorisation and Restriction of Chemicals, 2006 O.J. (L 396) 1 (EC).
[†] For brevity, we use "PBTs" and "CMRs" as shorthand for "PBT chemicals" and "CMR chemicals."

mathematical models that do not account for all of the complexities of real-world environmental conditions.

Scientifically, persistence, bioaccumulation, and toxicity are each continuous, multidimensional, and independently tested constructs. (Continuous constructs are those that can have any numerical value; multidimensional constructs do not exist along a single line.) For example, toxicity consists of multiple potential harms (e.g., neurotoxicity, immunotoxicity, and reproductive toxicity). Thus, from a scientific perspective, it is somewhat arbitrary to divide the world of chemicals into two bins: PBTs and non-PBTs.

Depending on the results of an experiment or modeling exercise, one or more numerical values may be used to describe the persistence, bioaccumulation, and toxicity of each chemical. The presentation of science for policy purposes is simplified when persistence, bioaccumulation, and toxicity determinations are made because a multidimensional, continuous variable is transformed into a one-dimensional, dichotomous variable. This simplification typically occurs when a "cutoff" value is applied to a particular test result (e.g., a chemical may be considered positive for persistence if its half-life in water is greater than or equal to two months). Since there is no single or universally accepted method for determining cutoff values for these properties, PBT determinations are not based exclusively on science but instead are hybrid science-policy decisions based partly on conventions that have evolved over time and partly on the number of chemicals that are judged to be manageable for risk assessment and/or management.

In this chapter, we explore these issues by tracing the regulatory origins of the PBT concept, developing a typology to categorize PBT policies for comparative study, and more fully explaining the purposes, scope, and methodological approaches of this book.

II. The significance and regulatory origins of the PBT concept

Persistence, bioaccumulation, and toxicity have long been appreciated as distinct concepts in the realm of science, but the grouping of the three properties into a term of art appears to have evolved more from policy innovation than from scientific discovery. Scientific literature began linking the concepts of persistence and accumulation of chemicals in the body beginning around the early 1970s (Peakall and Lincer 1970). The earliest mentions of PBT as a term of art, however, appear in the peer-reviewed scientific literature in the mid-to-late 1990s, about 20 years after government regulators had begun to use the term (Holm 1995; Brady and Patenaude 1996; Chapman et al. 1996; Pilgrim and Schroeder 1997). Notably, the term *persistent organic pollutant* (POP) predates the use of PBT as a term of art (Hutzinger, Sundström, and Safe 1976). When policy makers first became concerned about chemicals that persist or bioaccumulate in the environment, the term *PBT* was not in use, and thus the tendency was to interpret the concerns on a case-by-case basis. Several early events drew attention to the public health and environmental significance of chemical pollution.

Arguably, the first publicized cause for concern from chemical accumulation in food webs was the widespread incidence of methylmercury poisoning from the ingestion of fish in Japan in the late 1950s and throughout the 1960s (Wilson and Kazmierczak 2007). Since the cases occurred near Minamata Bay in southern Japan, the ailment became known as "Minamata disease." Individuals developed neurological disorders after consuming fish that contained elevated levels of methylmercury, the organic form of mercury that can bioaccumulate. Although several reports associate the Minamata incident with POPs, this singular event did not itself directly lead Japanese lawmakers to develop the PBT concept.*

Rachel Carson's *Silent Spring* was released in 1962 as Minamata disease was making headlines in Japan. Through her book, Carson (1962) stimulated greater global attention to the potential environmental implications of the widespread use of synthetic chemicals. Chlorinated chemicals had shown great promise as pesticides as early as the 1930s, and dichlorodiphenyltrichloroethane (DDT) was first used as a pesticide after World War II (Jensen 1972; Lipnick and Muir 2000; Rosenberg 2004; Selin 2010; Kinkela 2011). Paul Hermann Müller won the 1948 Nobel Prize in Medicine for discovering (in 1939) DDT's effectiveness against insects and household pests, which are vectors for typhus, malaria, typhoid fever, and cholera. The chemical grew in popularity because of its effectiveness at exposure levels too small to have a noticeable effect on the short-term health of humans. However, *Silent Spring* drew attention to the hazards of DDT and its derivatives or breakdown products (including dichlorodiphenyldichloroethylene [DDE]), which were shown to have had detrimental impacts on populations of predatory birds by causing eggshell thinning (Lincer 1975). DDT is currently banned in many countries but is still in use for malaria control in Africa, India, and other tropical areas.

When DDT was widely used as a pesticide, it was able to make its way up the food chain because of its persistence and tendency to bioaccumulate. It has a high chemical stability and a slow rate of transformation and degradation. DDT is therefore prone to persist in the environment, making it more likely to find its way into food webs. It is also lipid soluble and

---

* Japan responded by enacting effluent controls in its Water Pollution Control Law of 1970 (Powell 1991; MOE 2002).

not rapidly metabolized, meaning that it will bioaccumulate in the fats of animals that ingest it.

Several flame-retardant chemicals, including polychlorinated biphenyls (PCBs) and polybrominated biphenyls (PBBs), also exhibit these characteristics. In fact, the bioaccumulative nature of PCBs was discovered by accident in 1966 by the Swedish chemist Sören Jensen. During an investigation of the occurrence of DDT in humans and wildlife, Jensen studied the remains of an eagle found dead in the Stockholm archipelago. The eagle's tissue contained large amounts of unknown substances, which Jensen later identified as PCBs (Jensen 1972). PCBs are nonflammable and resistant to chemical and biological degradation. Thus, they were used widely in polyvinyl chloride (PVC) and paints to increase chemical resistance and as electrical insulators in capacitors and transformers.

In 1970, epidemiologists from the US Centers for Disease Control and Prevention (CDC) were called to investigate an unexplained neurological illness that had stricken three children of the Huckleby family and a dozen of their hogs in New Mexico (Roueché 1981). Mr. Huckleby and several other farmers had been feeding their hogs grain that was treated with a methylmercury fungicide. The family had been eating one of the hogs. By happenstance, the director of the CDC's Epidemiology Program had recently read a reprint of a ten-year-old article on Minamata disease. The Minamata report and a similar report documenting a cluster of poisonings in Guatemala led CDC epidemiologists to conclude that the Huckleby hogs and children had been poisoned by methylmercury that had accumulated in their tissues.

Follow-up investigations led the CDC to draw connections between cases of methylmercury poisoning in Minamata, Japan (1960), Iraq (1961), Pakistan (1963), Honshu, Japan (1965), Guatemala (1966), and New Mexico (1970). After the CDC presented its case to federal agencies in February 1970, the US Department of Agriculture suspended all federal registrations for products containing methylmercury that are labeled for use as seed treatments because they constituted an "imminent hazard to the public health" (Roueché 1981, p. 252).

Throughout the late 1960s and the early 1970s, human and animal exposures to substances, including methylmercury, chlorinated pesticides, and flame retardants, caused many high-profile incidents around the world. The result was a growing public demand for regulation (Nixon 1971; CEQ 1973; Grossman 2009). Indeed, throughout the early 1970s, public pressure driven by environmental and health advocates as well as the mass media led governments in industrialized nations to begin the challenging task of modernizing regulatory programs for chemical hazards.

As regulators developed policies to control potentially dangerous exposures to these substances, they took note of these three properties:

Some chemicals tended to persist in the environment, bioaccumulate in animals and/or humans, and presented toxicological hazards—hence PBT (CEQ 1970, 1971, 1977; Peakall and Lincer 1970; Boyd 2012). The regulatory history of PBTs therefore seems to have been borne out of this global yet disjointed effort to regulate chemical hazards that began in the early 1970s in Asia, Europe, and North America. For a list of these initial laws and the laws, agreements, and directives that followed, see Table 1.1.

Although the exact origins of the PBT concept are not entirely clear, Japan's 1973 Chemical Substances Control Law (CSCL) appears to be the first regulatory scheme to formally highlight persistence, bioaccumulation, and toxicity together in a common frame (Toda 2009; Uyesato et al. 2013; METI 2014). Annual reports from the US Council on Environmental Quality beginning in 1970, directed special concern toward chemicals (particularly DDT and PCBs) that persist in the environment and accumulate in the body, but the reports do not use the frame "PBT." The US Environmental Protection Agency (US EPA) issued regulations in 1972 to limit the use of DDT, citing persistence, bioaccumulation, long-range

Table 1.1 Significant PBT-related legislation, directives, regulations, and agreements

Policy name	Jurisdiction	Year enacted
Chemical Substances Control Law	Japan	1973
Council Directive on Pollution Caused by Certain Dangerous Substances Discharged into the Aquatic Environment of the Community	European Economic Community	1976
Toxic Substances Control Act	United States	1976
Great Lakes Water Quality Agreement	Canada and United States	1978
Convention on Long-range Transboundary Air Pollution	United Nations	1979
Canadian Environmental Protection Act of 1988	Canada	1988
Candidate Substances List for Bans and Phase-Outs	Ontario, Canada	1992
Design for the Environment Program	United States	1992
Accelerated Reduction/Elimination of Toxics	Canada	1994
Environmental Management Provisions on the First Import of Chemicals and the Import and Export of Toxic Chemicals	China	1994

(Continued)

Table 1.1 (Continued) Significant PBT-related legislation, directives, regulations, and agreements

Policy name	Jurisdiction	Year enacted
Sound Management of Chemicals Initiative	Commission of Environmental Cooperation of North America	1995
Toxic Substances Management Policy	Canada	1995
Integrated Pollution Prevention and Control Directive	EU	1996
Great Lakes Binational Toxics Strategy	Canada and United States	1997
Aarhus Protocol on Persistent Organic Pollutants	United Nations	1998
National PBT Strategy (Working Draft)	United States	1998
Oslo-Paris Convention for the Protection of the Marine Environment of the North-East Atlantic	United Nations	1998
Waste Minimization Program	United States	1998
Canadian Environmental Protection Act of 1999	Canada	1999
Toxics Release Inventory PBT Rule	United States	1999
Stockholm Convention on Persistent Organic Pollutants	United Nations	2001
Chemicals Management Plan	Canada	2006
PBT Rule	Washington State, United States	2006
Registration, Evaluation, Authorization, and Restriction of Chemicals	EU	2006
Strategic Approach to International Chemicals Management	United Nations	2006
Priority Persistent Pollutant List	Oregon, United States	2007
Toxics Reduction Act	Ontario, Canada	2009
Great Lakes Water Quality Protocol	Canada and United States	2012
Measures for the Registration of Hazardous Chemicals for Environmental Management	China	2012
Safer Consumer Products Regulations	California, United States	2013
Twelfth Five-Year Plan for Chemical Environmental Risk Prevention and Control	China	2013

transport, and toxicity in its rationale.* Moreover, both houses of Congress voted in favor of the Toxic Substances Control Act (TSCA) in 1972, but differences in the bills were not addressed before the session adjourned, and thus, no bill was enacted that year. Although the early US bills mentioned persistence, bioaccumulation, and toxicity in reference to particular substances (e.g., DDT), they did not combine the characteristics to identify a class of chemicals as Japan's law did. However, accounts indicate that the Japanese lawmakers were aware of these developments in the United States as they were developing the CSCL.

The Japanese government enacted the CSCL largely in response to a mass PCB poisoning incident in Kyushu in 1968. Hundreds of people contracted what became known as "Kanemi Yusho disease" or "Kanemi Rice Oil disease" after ingesting rice oil that was contaminated with PCBs (Umeda 1972; Star 1976; Yoshimura 2003; Stedeford and Banasik 2009). Scientists quickly associated PCB poisoning with severe skin ailments and, over time, noted similar skin disorders in children of exposed mothers and an increased incidence of cancer in exposed men (Selin 2010). The CSCL (still in effect) provides authority to ban the production, importation, and use of substances exhibiting all three characteristics—initially only PCBs in 1974, followed by hexachlorobenzene (HCB) in 1979, and so on (Shibata and Takasuga 2007).† The rest of the world, to varying degrees and through varying methods, has followed Japan (although not necessarily deliberately) in targeting certain PBTs for regulation.

PBT regulations in Europe developed in much the same way. In 1969, for example, a chemical plant in the former Federal Republic of Germany released a large amount of the insecticide endosulfan into the lower Rhine, which resulted in a massive, highly publicized fish kill (Greve and Wit 1971; van Urk, Kerkum, and van Leeuwen 1993). Even before that event, Sweden had been at the forefront of chemical regulation in Europe (Löfstedt 2003; Karlsson 2006; Selin 2010). The substitution principle, which calls for replacement of one substance with another substance, product, or process to achieve enhanced safety, first appeared in Sweden's 1949 Notification on Occupational Health and was codified in 1973 by the Act on Products Hazardous to Health and the Environment (Löfstedt 2003; Karlsson 2006; Selin 2010). The 1973 Act was Sweden's comprehensive response to growing concerns regarding PCBs, dioxin, and DDT, and it was also the first legislation that reversed the burden of proof of safety from government to industry (Briand 2010). The Act applies to potentially

---

* Consolidated DDT Hearings: Opinion and Order of the Administrator, 37 Fed. Reg. 13,369, 13,370–71 (July 7, 1972) ("Persistence and biomagnification in the food chain are, of themselves, a cause for concern, given the unknown and possibly forever undeterminable long-range effects of DDT in man, and the environment.")
† The Japanese government banned DDT and several other PBTs for use as pesticides before 1973.

hazardous substances, but it does not explicitly mention persistent or bio-accumulative chemicals. A report that accompanied the 1973 Act's draft legislation reveals that one of the motivating factors behind the legislation was the realization that certain chemicals, like PCBs, exhibit properties of both persistence and bioaccumulation.*

The EU adopted a 1976 directive to address discharges of chemicals into waterways, including specifically chemicals exhibiting PBT proper-ties.[†] Ultimately, the directive set limit values and quality objectives for 17 substances (including mercury and cadmium) out of 132 candidates.

The United States enacted the TSCA[‡] in 1976, largely in response to a rash of PCB and PBB incidents. In 1973, PBBs from the Michigan Chemical Corporation were accidentally introduced into animal feed and eventu-ally permeated much of Michigan's food chain. The state was compelled to destroy nearly 30,000 cattle, 6000 pigs, 1.5 million chickens, almost 900 tons of animal feed, 9 tons of cheese, 1.5 tons of butter, 17 tons of dry milk products, and almost 5 million eggs, at a total cost of over $200 million (Reich 1983). Similar incidents of contamination from PCBs and dioxin (2,3,7,8-tetrachlorodibenzo-p-dioxin), which was a contaminant in the herbicide 2,4,5-trichlorophenoxyacetic acid (e.g., in Agent Orange) and is chemically similar to PCBs, occurred in Connecticut, Pennsylvania, Tennessee, Washington state, and New York. The state of New York ulti-mately banned fishing in the upper Hudson River in 1975 because of PCB contamination (Boyle 1975; Star 1976). One of the most prominent incidents of improper disposal of substances containing dioxin and other chemicals occurred at the Love Canal neighborhood of Niagara Falls, New York. The Love Canal incident made national headlines in the late 1970s and helped motivate Congress to enact the Comprehensive Emergency Response, Compensation, and Liability Act of 1980 (Selin 2010). In 1978, illegal dump-ing of more than 30,000 gallons of PCB-contaminated oil along a 200-mile stretch of North Carolina highway led to a political battle over the dis-posal of contaminated soil that helped spawn the environmental justice movement in the United States (Geiser and Waneck 1994).

Also, in 1978, officials in Canada and the United States encouraged action to "virtually eliminate" PBTs from discharges into the Great Lakes basin by adopting the nonbinding Great Lakes Water Quality Agreement. By the 1990s, Canada and the United States had enacted a variety of regu-lations aimed at reducing risk of harm from selected PBTs in the Great Lakes basin and elsewhere (Davies 2006).

---

* Proposition [Prop.] 1973:17 Med förslag till lag om hälso- och miljöfarliga varor [government bill] (Swed.) (Governmental Bill 1973:17 on proposal for an act on products hazardous to health and the environment).

[†] Council Directive 76/464/EEC, 1976 O.J. (L129) (water pollution by discharges of certain dangerous substances).

[‡] Toxic Substances Control Act, 15 U.S.C. §§ 2601–2697 (2012) (hereinafter TSCA).

International efforts, primarily under the purview of the United Nations Economic Commission for Europe (UNECE) and the United Nations Environment Programme (UNEP), have also been significant. As scientists discovered PCBs and other chemicals in Arctic wildlife and indigenous Inuit populations in the 1980s, authorities around the world began to realize that some chemicals have the capacity to travel great distances and cross international borders (Selin 2010). The UNECE Convention on Long-Range Transboundary Air Pollution (CLRTAP), which entered into force in 1979, laid the groundwork for future controls on PBTs that pose cross-boundary risks.* The CLRTAP's 1998 Aarhus Protocol on POPs[†] is the precursor to the Stockholm Convention on POPs,[‡] which provides nations with a means for limiting exposures from PBTs that are capable of being transported long distances. Table 1.2 lists the original "dirty dozen" POPs and their uses. The Arctic Monitoring and Assessment Programme continues to show that animals at the top of marine Arctic food webs carry high levels of many of these POPs, although the concentrations are slowly declining (Selin 2010).

In summary, although use of the term *PBT* in regulatory policy appears to have begun in Japan, it was soon after used in other jurisdictions. Once persistence, bioaccumulation, and toxicity were recognized as a potentially troublesome combination of properties through the early efforts to regulate methylmercury, DDT, dioxins, and PCBs, governments began to look for this combination of properties in their search for chemical hazards that might require regulation.

As discussed later in Chapter 3, there are potentially hundreds of substances in commerce with PBT properties. An important regulatory goal of many of the programs that we discuss below is to encourage producers to innovate by creating chemicals without these properties. In some cases, however, PBT attributes may be indicators of commercially valuable, often necessary, properties. It is therefore unlikely that PBTs will disappear entirely.

Indeed, PBTs are used as heavy-duty lubricants, water repellents, antioxidants and stabilizers in plastics, dyes and pigments, fire retardants, solvents, pesticides, and herbicides and are in many pharmaceuticals and cosmetics. Water repellents, for example, must include compounds that are hydrophobic and lipophilic. These are the properties that repel water; however, they are also the properties that contribute to bioaccumulation. Outdoor paints must contain persistent compounds that resist degradation and corrosion over time, and pigments in colors must be persistent

* Convention on Long-Range Transboundary Air Pollution, November 13, 1979, 1302 U.N.T.S. 217.
† Protocol to the 1979 Convention on Long-Range Transboundary Air Pollution on Persistent Organic Pollutants, June 24, 1998, 2230 U.N.T.S. 79.
‡ Stockholm Convention on Persistent Organic Pollutants, May 22, 2001, 40 I.L.M. 532 (entered into force May 17, 2004).

Table 1.2 The original 12 chemicals (the "Dirty Dozen") listed under the Stockholm Convention on POPs[a]

Name	Uses
Aldrin	Classified as a pesticide; applied to soils to kill various insect pests.
Chlordane	Classified as a pesticide; used extensively to control termites and as a broad-spectrum insecticide on a range of agricultural crops.
DDT	Classified as a pesticide; used widely to combat malaria, typhus, and other diseases spread by insects. Also sprayed on a variety of agricultural crops, especially cotton. DDT continues to be applied against mosquitoes in several countries to control malaria.
Dieldrin	Classified as a pesticide and a degradation product of aldrin; used to control termites and various other pests.
Endrin	Classified as a pesticide; sprayed on crops such as cotton and grains to control insects. It is also used to control rodents.
Heptachlor	Classified as a pesticide; primarily used to kill soil insects and termites. Heptachlor has also been used to kill cotton insects, grasshoppers, and other pests, including mosquitoes.
HCB	Classified as an industrial chemical, pesticide, and by-product; first introduced in 1945 to treat seeds. HCB kills fungi that affect food crops. It was widely used to control wheat bunt and is a by-product of the manufacture of certain industrial chemicals.
Mirex	Classified as a pesticide; used mainly to combat fire ants, and some other types of ants and termites. Also used as a fire retardant in various products.
Toxaphene	Classified as a pesticide; used on cotton, cereal grains, fruits, nuts, and vegetables. It has also been used to control ticks and mites in livestock.
PCBs	Classified as an industrial chemical; used in industry as heat exchange fluids, in electric transformers and capacitors, and as additives in various products.
Polychlorinated dibenzo-*p*-dioxins (PCDDs)	Classified as a by-product; produced unintentionally due to incomplete combustion, as well during the manufacture of pesticides and other chlorinated substances.
Polychlorinated dibenzofurans (PCDFs)	Classified as a by-product; produced unintentionally from many of the same processes that produce dioxins, and also during the production of PCBs.

[a] Additional details on each chemical are available from the Stockholm Convention website (http://chm.pops.int/Home/tabid/2121/Default.aspx) and the POPs Toolkit (http://www.popstoolkit.com/about.aspx).

to avoid fading. Similarly, many pesticides and herbicides must persist on crops and in soils throughout the growing season to effectively destroy or repel pests and weeds. In short, persistence and bioaccumulation properties can have significant economic utility in some applications.

Many PBTs are chlorinated compounds. Chlorine is a good example of a chemical for which the very properties that make it useful in certain products may simultaneously make it worrisome. The US EPA has observed that "the characteristics that make chlorine a superb cleaning/ bleaching agent also contribute to its adverse impact on surrounding environments when released from the production process" (Mansfield and Depro 2000, pp. 2–14). In many applications, the risks of using these chemicals may outweigh their economic and safety benefits. However, it is recognized that the public health benefits of DDT use in controlling the spread of malaria in sub-Saharan African nations may outweigh its risks to the environment, hence the exceptions written into the Stockholm Convention (Rosenberg 2004). Other chlorinated chemicals—and the chlorine industry in general—have significant benefits. Chlorine itself is used as a disinfecting agent to make water safe to drink. Chlorine compounds are also key constituents in household disinfectants and bleaches and are vital components in many crop protection chemicals and in the production of pharmaceuticals (Mansfield and Depro 2000; WCC 2002; ACC 2012, 2014; EuroChlor 2012; C4 2014; LeChevllier and Au 2014). Other distinct uses of chlorine and chlorine compounds are in the production of ultrapure silicon, the primary material of photovoltaic power cells; polyaramide fibers, which replaced asbestos in certain applications and reinforce fiber optic cables for telecommunications; and silicon chips in microprocessors (WCC 2002).

Of course, chlorine itself is not a PBT, and not all chlorinated chemicals are PBTs. Nonetheless, chlorinated chemicals provide a good example of simultaneous hazard and economic benefit, often from the same or related properties. They therefore demonstrate the complexities of PBT management—that it is worthwhile to distinguish high- from low-benefit and high- from low-risk uses and that an outright ban of a PBT is not always a wise course of action.* Thus, regulatory frameworks should develop mechanisms to meaningfully incorporate information on the potential for exposure from different uses when weighing risk management options.

III.   Typology of PBT policies

As we survey the range of policies that make use of the PBT determination in this book (Chapters 4, 5, and 6), we find it useful to draw a conceptual

---

* For a discussion of the economics of chemical substitution, see Åström et al. (2013) and Friege (2013).

distinction between three kinds of policies: (A) those that seek to elevate the priority for risk assessment given to PBTs by regulators and industry, (B) those that impose some form of informational requirement on industry, as a condition of continued commercial use of a PBT, and (C) those that directly discourage, restrict, or prohibit the use of PBTs in commerce. Our focus is on how PBT policies influence the use and management of existing industrial chemicals, although we also note some ways that the PBT determination is used in the regulatory evaluation of new chemicals.

A. Priority setting

Policy makers are faced with limited resources (e.g., money, staff, scientific expertise, and time) to allocate to chemical risk management. Risk assessment generally includes assessments of hazard properties, dose–response relationships, potential for exposure, and risk characterization. As part of or as a precursor to risk assessment, chemicals can undergo a PBT assessment to determine whether they meet or are likely meet standards (e.g., cutoff thresholds) for each of the three properties. PBT assessment may be used as one mechanism to guide regulators and industrial managers in making resource allocation decisions by identifying which chemicals or groups of chemicals should be prioritized over others for risk assessment and management.

When the amount of resources available for risk management is fixed, the practical effect of the PBT determination is to increase the allocation of resources to PBTs while reducing the allocation of resources to non-PBTs. Alternatively, the PBT determination may be seen as an effort to persuade policy makers to increase the total amount of resources available for chemical risk management, thereby giving greater priority in society to chemical risk management. In either case, the PBT determination acts to influence priority setting in the allocation of scarce resources.

The Canadian Environmental Protection Act (CEPA) of 1999 and the Chemicals Management Plan (discussed in Chapter 5) together provide the clearest example of the PBT concept appropriated as a priority-setting tool. CEPA 1999 mandated that 23,000 substances in Canadian commerce be categorized and prioritized for screening level risk assessments by 2006. The legislation used persistence, bioaccumulation, and toxicity criteria (among others) to prioritize the substances for assessment.

The state of Washington in the United States has its own PBT program (described in Chapter 6). Washington's Department of Ecology has generated an official list of PBTs, but the list does not trigger any legally binding obligations on government or industry. Instead, the list of PBTs indicates that resources for risk management may be necessary and appropriate. In effect, the list is used as a priority-setting device to encourage the development of cooperative risk management plans by government and industry.

In contrast to Washington's rather simple priority-setting scheme, imagine a numerical system where each industrial chemical is scored for its various properties, with a higher (lower) score on each property suggesting the potential for larger (smaller) degrees of risk. Regulators may decide to give higher priority to chemicals with high scores than to chemicals with low scores. The US EPA uses this kind of numerical scoring system (described in Chapter 5) to help allocate the agency's resources to industrial chemicals of concern. Scoring systems may go beyond chemical characteristics and include potential for release and exposure.* For such scoring systems, a central question is what weight should be given to persistence, bioaccumulation, and toxicity and to the combination of the three properties?

B.   *Information requirements*

Policy makers sometimes use the PBT determination to impose legally binding informational requirements on industry. The resulting information may foster enhanced worker or consumer safety or may be used in the marketplace to supply an advantage for greener products and chemicals.

A PBT determination may trigger changes to safety data sheets, key information that is shared in the supply chain. Safety data sheets may contain precise handling practices to ensure safety and/or procedures to follow in the case of an accidental release.

Under the European REACH Regulation, registration dossiers for PBTs have special requirements for use-specific exposure scenarios and risk characterization, potentially providing a competitive cost advantage to registering chemicals that are not PBTs. By requiring data on potential exposure scenarios and risk management measures for specific uses, REACH also sets the stage for the risk–benefit analysis of PBTs that is permitted under the authorization procedure (described in Chapter 4).

Another information requirement, aimed at specific industrial facilities that release PBTs, is to require them to publicly report emissions or releases. As community leaders and residents realize that PBTs are being emitted into their local environment, they may exert pressure on facilities to control, mitigate, or prevent such emissions. The United States' Toxic Release Inventory (TRI) is an example of such a program, and research concludes that the TRI program has contributed to significant reductions in chemical emissions (Hamilton 2005; Kerret and Gray 2007).[†] This research also demonstrates that investors/shareholders respond to the

---

* For an illustration of an analytic prioritization scheme that includes information on exposure as well as persistence, bioaccumulation, and toxicity, see Dulio and von der Ohe (2013).

[†] However, see Kerret and Gray (2007), who find that it is difficult to distinguish the effects of pollution registries from other control activities.

publicly available TRI data, exerting pressures on companies through stock market signals (Hamilton 2005). The US EPA has imposed more stringent TRI reporting requirements for emitted chemicals that are classified as PBTs. Other nations have similar programs as well.

The obligation to spread information on PBT properties throughout the supply chain of a chemical may cause shifts in a chemical manufacturer's research, investment, and marketing decisions. Thus, informational policy instruments can have notable impacts on chemical selection, management, and any resultant risks to human health and the environment. One of the themes that continuously appears throughout this book is that additional research into the nature of PBTs (and other chemicals) can greatly improve risk management.

C.   Directly discourage, control, restrict, or prohibit use

Some PBT policies are designed to directly discourage, restrict, or prohibit industrial uses of PBTs. There are some 30 different policy tools that regulators can apply to manage risks from chemicals, including restrictions on the quantity of manufacture, sale, import, export, or use; on the amount, location, and conditions of releases; on labeling, handling, and storage; and on the generation and submission of information as mentioned above. The risk management process includes identifying PBTs, determining who (government or industry) has the burden of proving that the chemicals are used safely or with an acceptable threshold of risk, determining what risk management tool or mechanism to apply to reduce risk, and implementing risk management. Depending on the jurisdiction, these processes usually do not occur in an orderly sequential fashion, but rather they are overlapping and/or embedded practices in jurisdictions' overall chemical risk management frameworks. Here, we find it conceptually useful to describe each of these processes separately to give readers a clearer idea of all of the different ways in which the PBT concept is being deployed.

Risk management of PBTs begins with *identifying* which chemicals meet the criteria for PBT properties. Often, chemicals that meet the criteria for all three (or sometimes two) of the properties are included on lists. These lists also usually include additional categories of chemicals of concern (e.g., CMRs). Although such lists generally have a priority-setting purpose for government, their publication can also have a stigmatization effect that discourages companies from producing, distributing, or using the listed chemicals. In the marketplace, such lists may be seen as "grey lists" or "black lists"—even when it is lawful to produce, distribute, and sell chemicals on the list.

Since the PBT determination by itself does not reflect a risk assessment of specific uses, lawmakers may also design policies to ensure that specific uses of PBTs are analyzed with respect to risk to ensure that safety is accomplished through risk management. However, effective risk management of PBTs is challenging precisely because the substances are persistent and bioaccumulative once released into the environment. Depending on the results of risk assessment, the PBT determination may be seen as the front end of a risk management process that can lead to controls, restrictions, or prohibitions on one or more uses of a chemical.

The *burden of risk assessment and management* may be placed on government and/or industry. Chemical risk management necessarily involves cooperation and coordination among industry, government, public interest stakeholders, and even consumers. When it comes to initiating the regulatory process, legal frameworks have traditionally placed the burden on the government of demonstrating that a chemical use is unsafe. That is, once the government demonstrates that a chemical use does not meet a legal definition or standard of safety, the government is then authorized to apply risk management regulations. Recently, jurisdictions have begun transferring that burden onto industry. One of the cornerstones of REACH is the "no data, no market" principle—industry has both the burden of supplying data on risk and the burden of certifying that a chemical is used safely or it cannot market the chemical in Europe. The global trend is to place greater analytic burdens on industry to demonstrate proper risk management and safety.

Risk management can come in many forms. Many jurisdictions apply bans or phase-outs to PBTs, subject to use-specific exemptions. Under Japan's CSCL, chemicals that meet the criteria for all three PBT properties are banned. Europe's REACH program, on the other hand, calls for the phase-out of PBTs unless producers can demonstrate, in a use-specific authorization process, that the benefits of the PBT outweigh the risks and that there are no suitable substitutes. Canada's Prohibition of Certain Toxic Substances regulations include both lists of chemicals that are banned and those that are banned but subject to use-specific exemptions. In the United States, on the other hand, very few chemicals are formally banned. Rather, the use of chemicals in consumer products tends to be governed as much by state products liability laws as formal regulation under TSCA. The PBT concept is nonetheless important in several other federal laws in the United States as well as state regulatory systems.

In regulatory systems around the world, at all levels, and in the private marketplace, the PBT concept is a central component in identifying priority chemicals for assessment and management. Thus, it is quite apparent that the PBT determination plays an important role in stigmatizing, regulating, and prohibiting the production and/or use of industrial chemicals. In the chapters to come, we explore the subtle differences in how the PBT determination is used in different regulatory programs.

IV.   Purposes and scope of the book

We are particularly interested in the extent of international variation in how PBT determinations are made and in how regulators make use of PBT determinations. We first examine how PBT determinations are made by public agencies, over time and in different jurisdictions, and then explain the various uses of PBT determinations by decision makers in the public and private sectors. Our interest is in the PBT practices and policies, not the outcomes for any specific substances. On the basis of this examination of practices and policies, we offer suggestions for both improvement and harmonization.

Although the focus of this book is limited to PBTs, its relevance to chemical management is somewhat broader than this single category of chemicals. First, we consider POPs to be a subset of PBTs that are within the purview of our examination. Second, many of the assessment and management processes that we describe apply to more chemicals than those that fall into the PBT category. In particular, we present some findings and recommendations about what we call "partial PBTs"—chemicals that satisfy two of the three PBT properties. In most chemical management regimes, chemicals are subjected to PBT assessment to determine whether they will meet the legal criteria for being formally considered to be persistent, bioaccumulative, and/or toxic. We call substances that meet the criteria for all three properties PBTs, but a PBT assessment may also reveal that a chemical meets the criteria for only one or two of the properties rather than all three. The initial phases of PBT policies often apply to non-PBTs as well because it is only *after* PBT assessment that it is discovered that these chemicals are actually not PBTs after all.

Moreover, several regulatory regimes treat partial-PBTs that have two of the three properties differently than they do chemicals that have one or no PBT properties. REACH, for example, applies the designation "very persistent, very bioaccumulative" (vPvB) to chemicals that are highly persistent and bioaccumulative but for which toxicity is unknown or poorly characterized. Chemicals that are vPvB could therefore be described as partial PBTs, but we view them as a subset of PBTs because toxicity is generally implied for these chemicals. In identifying priority chemicals for risk assessment, Canada targeted not only potential PBTs but also chemicals that are potentially bioaccumulative and toxic or persistent and toxic. Japan also applies risk management to partial PBTs.

We concentrate primarily on industrial chemicals, recognizing that agricultural chemicals, biocides, and pharmaceuticals raise some different scientific and policy issues. Our primary focus is on PBT practices and policies in Asia, Europe, and North America, but we also address some other international activities. Although some PBT policies have arisen in the consideration of new chemicals, our focus is how PBT practices and policies should be integrated into the accelerated review of existing chemicals.

V. *Method*

As a framework for analysis, we start with some basic concepts and principles of risk management that are applicable to a wide range of technologies. For simplicity, we refer to these collectively as "management principles." The principles are drawn from basic texts, classic papers on risk assessment and management, reports from the Organization for Economic Cooperation and Development (OECD), and official statements of government policy.

Throughout the report, we explore whether PBT practices and policies are consistent with these management principles. In cases of deviation from the principles, we ask whether the deviation can be justified based on either practical considerations or distinctive aspects of PBTs that are different from other chemicals and other technologies. In cases where deviations do not seem justified, we suggest how PBT practices and policies might be better aligned with the principles.

We apply seven management principles:

1. *Information Quality.* The scientific information relied upon by industry and regulators should meet high standards of clarity and accuracy.* When scientific opinions are employed, the information and documentation that form the basis for such opinions should be made publicly available, except for cases of legitimate protection of confidentiality. Such opinions should be developed through a process that is transparent to stakeholders and allows for public participation. We call this the "information quality" principle, a key facet of evidence-based decision making.

    In the management of potentially hazardous technologies, scientific information has a wide variety of purposes: making official determinations of hazard, setting priorities for risk assessment and management, preparing exposure and risk assessments, preparing risk management, preparing impact assessments (e.g., cost–benefit and socioeconomic analyses), and evaluating alternative technologies (e.g., relative risk comparisons).

    Because of time constraints in government and industry, scientists may err by not fully adhering to the information quality principle. Illustrations of the principle include transparency (e.g., about experimental design, data collection methods, and modeling), validity and

---

* A concise statement of quality of information principles can be found in the Canadian Privy Council Office (2003).

reliability of research findings, disclosure of assumptions, consideration of all relevant and credible data, disclosure of limitations and uncertainties, and peer review by qualified experts.

In PBT assessment, information quality issues are a key concern, and clarity is needed as to which parties (e.g., government and/ or industry) are responsible for meeting information quality standards. The same information quality standards should be applicable regardless of who is producing information about PBTs.

2. *Precaution.* When there is uncertain evidence of a potential threat to human health, safety, and the environment, lack of full scientific evidence of adverse effects should not be used as an excuse to postpone the implementation of cost-effective protective measures.* This is a version of the precautionary principle.[†]

In light of the historical disputes about the need for regulation of toxic substances such as tobacco, asbestos, and lead, policy makers now recognize that regulation of a potentially hazardous technology may be prudent even though all of the scientific issues about causes and effects are not fully resolved. The precautionary principle— sometimes called the precautionary approach in North America— was articulated to capture this sentiment (EEA 2013).

The purpose of identifying PBTs is to help risk managers in industry and government implement precautionary measures that will protect human health and the environment. A wide range of precautionary steps are possible such as gathering more data to clarify the extent of releases, potential exposures and risk, reducing releases of pollutants into the environment through controls, adding appropriate warning labels for workers and/or consumers, or completely replacing a PBT with a safer alternative (European Commission 2000). The steps are considered precautionary rather than preventive because, at the point when a chemical is regulated, it is often not feasible to demonstrate with scientific certainty that any particular use of a PBT is causally linked with particular damages to human health or the environment.

3. *Priority Setting.* Since scarce resources typically preclude the elimination of all risks from technologies, decision makers should establish a systematic priority-setting process to ensure that the most undesirable risks are given appropriate priority, in light of the benefits and potential for risk reduction (Fosler 1995; Davies 1996; OECD 2010). We call this the "priority setting" principle.

---

* UN Conference on Environment and Development, June 3–14, 1992, Rio Declaration on Environment and Development, princ. 15, UN Doc. A/CONF.151/5/Rev.1 (June 14, 1992).
[†] This is one formulation of the precautionary principle (see Løkke 2006).

Government and industry lack sufficient resources to identify and regulate all technologies that may pose some risk to public health and the environment. In the face of limited resources, priorities should be set based on procedures that target opportunities for maximum risk reduction.

In Asia, Europe, and North America, tens of thousands of industrial chemicals are now in use. Guidance is needed as to which chemicals should be evaluated first and which can be deferred for later evaluation. Historically, priority-setting schemes have emphasized CMRs and/or those chemicals that are widely used in ways that could lead to human and environmental exposure. A key question is whether PBT itself is a meaningful category, whether the three properties together represent a categorically greater concern relative to any of the individual characteristics alone, and how much priority should be assigned to PBTs relative to hazard classifications and other priority-setting factors (e.g., production volume and dispersive uses).

4. *Exposure and Risk Assessment.* Risk is defined by the presence of a hazard and exposure to the hazard. Before making risk management decisions about a specific technology, decision makers should be informed of a technology's potential for exposure and risk in addition to whether or not the technology has the potential to be hazardous. We call this the "exposure and risk assessment" principle (Wilson and Crouch 2001; OECD 2010).

In the United States, there were efforts in the 1970s to regulate carcinogens in the workplace based on a qualitative hazard determination (e.g., demonstrated carcinogenicity in laboratory animals or in epidemiological studies of exposed workers), without any quantitative assessment of a worker's exposure pattern or risk of disease. In 1980, in a case involving benzene exposure, a divided US Supreme Court held that the US Department of Labor's Occupational Safety and Health Administration is required to demonstrate that the level of risk to a worker is significant before issuing regulations to reduce the risk (Graham, Green, and Roberts 1988).*

The exposure and risk assessment principle is not necessarily in conflict with the precautionary principle. Even in a situation where the precautionary principle is being invoked, the European Commission (2000) has recognized the importance of undertaking a risk assessment to inform regulatory decisions. And the US Supreme Court indicated in the benzene case that the significant-risk requirement is not a mathematical straightjacket that forbids precautionary action.

---

* *Industrial Union Department v. American Petroleum Institute*, 448 U.S. 607 (1980).

A key policy question we address is whether PBTs should be regulated based only on a qualitative determination of potential hazard or whether indications of release, exposure, and risk should be required. For PBT policies to satisfy the exposure and risk assessment principle, regulators need to buttress PBT determinations with some information drawn from the tools of exposure and risk assessment.

A related policy question we address is whether certain uses of PBTs should be permissible if industry supplies credible evidence that potential releases are adequately controlled and/or exposures will be too small or infrequent to pose a risk to human health and the environment. In the United States, such risk-based approaches tend to be applied to all chemicals, accounting for properties including persistence, bioaccumulation, and toxicity, as well as others. Under Europe's REACH regulation, however, PBTs seem to be regulated as a separate category (somewhat like carcinogens), on the assumption that risk and safety determinations cannot be made. Risk assessments for PBTs may be more complicated than risk assessment of some non-PBTs because of the potential for cumulative exposures: Once a PBT is released, slow degradation provides for increasing environmental concentrations that, over time, make more of the chemical available for bioaccumulation. And such concerns are amplified when PBTs are used widely in ways that trigger significant and repeated releases into the environment. Despite the added complexity of risk assessment of PBTs, US regulators have a long history of applying risk assessment to substances such as mercury, lead, dioxin, PCBs, and other chemicals that are persistent, bioaccumulative, and toxic.

5. *Differentiation.* If the same technology is likely to have different risks in different real-world applications, the risk management and regulatory processes should be discerning enough to distinguish low-risk from high-risk applications (as well low-benefit versus high-benefit applications). We call this the "differentiation" principle.

In the 1980s, the US EPA attempted to ban virtually all industrial uses of asbestos, even though some uses posed significantly less risk than others did. A federal appeals court overruled US EPA's action, holding that, under TSCA, each use of asbestos must be considered from the perspective of risk, benefit, and cost.* The European Commission, when considering restrictions on the use of industrial chemicals, also analyzes toxic chemicals on a use-by-use basis, although typically with socioeconomic analysis rather than the cost–benefit analysis performed in the US.

---

* *Corrosion Proof Fittings v. Environmental Protection Agency,* 947 F.2d 1201 (5th Cir. 1991).

For PBTs, a key question is how different uses, alone and in combination, contribute to releases, environmental loadings, exposures, and risks and how regulatory regimes can be designed to be flexible enough to differentiate regulatory stringency based on this use information. In Europe, for example, some degree of use-by-use differentiation is already envisioned under the REACH authorization and restriction procedures, but REACH also calls for consideration of the cumulative impacts of multiple uses.

6. *Rational Alternatives.* When innovative substitutes for a potentially hazardous technology are considered, the anticipated risks, costs, and benefits of the substitute technologies should also be considered by decision makers (OECD 2010; Sunstein 2002). We call this the "rational alternatives" principle.

   One of the principal challenges in technological risk management is to ensure that the process of innovation leads to net benefits and, where benefits are disregarded, a net reduction in risk to human health and the environment rather than simple replacement of existing risks with a new portfolio of risks (Graham and Wiener 1997). The emerging field of green and sustainable chemistry is intended to channel innovative energies in the direction of safer chemicals and more sustainable chemistry-based products (Anastas and Warner 2000).

   With respect to PBTs, the objective is to find non-PBT (or other) alternatives that can serve the same function as the PBT at reasonable cost, without generating unacceptable (and/or different) risks to human health and the environment. This comparative risk tradeoff analysis is particularly difficult when relatively little data are available for the substitute technology or chemical.

7. *Value of Information.* When an in-depth evaluation of a technology is undertaken after a screening exercise, the initial determination may need to be revised based on more definitive information (Canadian Privy Council Office 2003). Since the collection of new scientific information is costly (in time and money), research investments should focus on projects where the value of additional information justifies the costs (NRC 2009). We call this the "value of information" principle.

   When uncertainties about the risk of a specific technology can be clarified or eliminated with targeted research activities, investments in additional research can be quite valuable. When the cost of research is small compared to the possible damages from an ill-informed risk management decision, the case for additional research investment is particularly strong. On the other hand, the lack of established "stopping criteria" for research activity can create a syndrome of indefinite study without decisive action. Delay of regulatory decision making generates costs and risks of its own. An analytic tool called value-of-information analysis was developed by decision theorists to help define stopping rules and optimal investments in research (Raiffa 1970; Clemen 1997; Yokota and Thompson 2004; Boardman et al. 2006).

Since there are substantial uncertainties associated with the risks of PBTs and possible substitute chemicals, the value-of-information framework may shed light on how to make research investments and regulatory decisions in the face of uncertainties. Efficiency in information generation also entails coordination of the research and development efforts of government, academia, and industry as they relate to PBTs.

In later chapters, we pinpoint some areas where practices and policies appear to be consistent with these seven principles and other areas where deviations from the principles have occurred, appear to be occurring, and are likely to occur in the future. Although we have found it useful to compare current PBT determinations and policies to these principles, much of our analysis and many of our suggestions do not hinge on whether the reader concurs with all of the seven management principles.

VI.   Organization of the book

Following this introductory chapter, Chapter 2 examines the role of scientific data, modeling, and conventions in the making of PBT determinations. We distinguish areas of stronger and weaker scientific consensus while highlighting sources of significant scientific uncertainty and judgment. Chapter 3 explores the number of PBTs in commerce and the growing role of weight of evidence (WOE) in the making of PBT determinations. Since several jurisdictions are moving from numeric bright lines to WOE approaches to making PBT determinations, we identify issues that are likely to arise in WOE judgments and suggest the need for some guidelines to enhance rigor, predictability, and credibility. Chapter 4 examines international and regional policies toward PBTs, where policies concern measures taken to protect public health and the environment from PBT exposures. Chapters 5 and 6 examine national and subnational PBT policies, respectively. Finally, Chapter 7 summarizes our major findings and recommendations and concludes with a case for greater harmonization of PBT determinations and policies. The Appendix provides a list of all interviewees, compensated peer reviewers, and commenters.

References

ACC (American Chemistry Council). 2012. "Chlorine Chemistry: Industry Fact Sheet," March. Available at http://www.americanchemistry.com/Jobs /EconomicStatistics/Chemistry-in-Economy/Chlorine-Chemistry-Industry -Fact-Sheet.pdf.

ACC (American Chemistry Council). 2014. "Chlorine Chemistry: Economic Benefits." Accessed September 2. Available at http://chlorine.american chemistry.com/Chlorine-Benefits/Economic-Benefits.

Anastas, P.T., and J.C. Warner. 2000. *Green Chemistry: Theory and Practice*. New York: Oxford University Press.

Åström, S., M. Lindblad, J. Westerdahl, and T. Rydberg. 2013. "Are Chemicals in Products Good or Bad for Society?—An Economic Perspective." In *Global Risk-Based Management of Chemical Additives II: Risk-Based Assessment and Management Strategies*, edited by B. Bilitewski, R.M. Darbra, and D. Barceló, 109–36. New York: Springer.

Boardman, A., A. Vining, D.L. Weimer, and D.H. Greenberg. 2006. *Cost–Benefit Analysis: Concepts and Practice*, 3rd ed. Upper Saddle River, NJ: Pearson/ Prentice Hall.

Boyd, W. 2012. "Genealogies of Risk: Searching for Safety, 1930s–1970s." *Ecology Law Quarterly* 39:895–987.

Boyle, R.H. 1975. "The Spreading Menace of PCB." *Sports Illustrated*, December 1. Available at http://www.si.com/vault/1975/12/01/614397/the-spreading -menace-of-pcb.

Brady, K.S., and L.M.F. Patenaude. 1996. "Canada's Life Cycle Activities and Product Policies." *International Journal of Life Cycle Assessment* 1:40–2.

Breidenbach, A.W., C.G. Gunnerson, F.K. Kawahara, J.J. Lichtenberg, and R.S. Green. 1967. "Chlorinated Hydrocarbon Pesticides in Major River Basins, 1957–65." *Public Health Reports* 82:139–56.

Briand, A. 2010. "Reverse Onus: An Effective and Efficient Risk Management Strategy for Chemical Regulation." *Canadian Public Administration* 53:489–508.

C4 (Canadian Chlorine Chemistry Council). 2014. "Chlorine Benefits." Accessed September 19. Available at http://www.cfour.org/chlorine/.

Canadian Privy Council Office. 2003. *A Framework for the Application of Precaution in Science-Based Decision Making About Risk*. Ottawa, ON, Canada: Privy Council Office.

Carson, R. 1962. *Silent Spring*. Cambridge, MA: Riverside Press.

CEQ (Council on Environmental Quality). 1970. *Environmental Quality: The First Annual Report of the Council on Environmental Quality*. Washington, DC: US Government Printing Office.

CEQ (Council on Environmental Quality). 1971. *Environmental Quality: The Second Annual Report of the Council on Environmental Quality*. Washington, DC: US Government Printing Office.

CEQ (Council on Environmental Quality). 1973. *Environmental Quality: The Fourth Annual Report of the Council on Environmental Quality*. Washington, DC: US Government Printing Office.

CEQ (Council on Environmental Quality). 1977. *Environmental Quality: The Eighth Annual Report of the Council on Environmental Quality*. Washington, DC: US Government Printing Office.

Chapman, P.M., I. Thornton, G. Persoone, C. Janssen, K. Godtfredsen, and M.N. Z'Graggen. 1996. "International Harmonization Related to Persistence and Bioavailability." *Human and Ecological Risk Assessment* 2:393–404.

Clemen, R.T. 1997. *Making Hard Decisions: An Introduction to Decision Analysis*, 2nd ed. Belmont, CA: Duxbury Press.

Davies, J.C., ed. 1996. *Comparing Environmental Risks: Tools for Setting Government Priorities*. Washington, DC: Resources for the Future.

Davies, K. 2006. "Strategies for Eliminating and Reducing Persistent Bioaccumulative Toxic Substances: Common Approaches, Emerging Trends, and Level of Success." *Journal of Environmental Health* 69:9–16.

Dulio, V., and P.C. von der Ohe, eds. 2013. *NORMAN Prioritisation Framework for Emerging Substances.* Verneuil en Halatte, France: NORMAN Association.

EEA (European Environment Agency). 2013. "Late Lessons from Early Warnings: Science, Precaution, Innovation." EEA Report No 1/2013, January 23. Available at http://www.eea.europa.eu/publications/late-lessons-2.

EuroChlor. 2012. "Chlorine Industry Review 2011–2012." Available at http://www.eurochlor.org/media/63146/2012-annualreview-final.pdf.

European Commission. 2000. "Communication from the Commission on the Precautionary Principle." COM/2000/0001/final.

Fosler, R.S. 1995. *Setting Priorities, Getting Results: A New Direction for EPA.* Washington, DC: National Academy of Public Administration.

Friege, H. 2013. "Two Sides of One Coin: Relations between Hazardous Substances and Valuable Resources." In *Global Risk-Based Management of Chemical Additives II: Risk-Based Assessment and Management Strategies,* edited by B. Bilitewski, R.M. Darbra, and D. Barceló, 155–69. New York: Springer.

Geiser, K., and G. Waneck. 1994. "PCBs and Warren County." In *Unequal Protection: Environmental Justice and Communities of Color,* edited by R.D. Bullard, 43–52. San Francisco: Sierra Club Books.

Graham, J.D., and J.B. Wiener, eds. 1997. *Risk versus Risk: Tradeoffs in Protecting Health and the Environment.* Cambridge, MA: Harvard University Press.

Graham, J.D., L.C. Green, and M.J. Roberts. 1988. *In Search of Safety: Chemicals and Cancer Risk.* Cambridge, MA: Harvard University Press.

Greve, P.A., and S.L. Wit. 1971. "Endosulfan in the Rhine River." *Journal of the Water Pollution Control Federation* 43:2338–48.

Grossman, E. 2009. *Chasing Molecules: Poisonous Products, Human Health, and the Promise of Green Chemistry.* Washington, DC: Island Press.

Hamilton, J.T. 2005. *Regulation through Revelation: The Origin, Politics, and Impacts of the Toxics Release Inventory Program.* New York: Cambridge University Press.

Heindel, J.J., R.T. Zoeller, S. Jobling, T. Iguchi, L. Vandenberg, and T.J. Woodruff. 2013. "What is Endocrine Disruption All About?" In *State of the Science of Endocrine Disrupting Chemicals—2012,* edited by Å. Bergman, J.J. Heindel, S. Jobling, K.A. Kidd, and R.T. Zoeller, 1–22. Geneva: World Health Organization and United Nations Environment Programme. Available at http://www.who.int/ceh/publications/endocrine/en/.

Holm, S.E. 1995. "Removal of Persistent Bioaccumulative Toxic Chemicals from Pulp and Paper Mill Effluent Streams." Paper presented at the National Meeting of the American Chemical Society, Chicago, August 20–25.

Hutzinger, O., G. Sundström, and S. Safe. 1976. "Environmental Chemistry of Flame Retardants Part I. Introduction and Principles." *Chemosphere* 5:3–10.

Jensen, S. 1972. "The PCB Story." *Ambio* 1:123–31.

Karlsson, M. 2006. "The Precautionary Principle, Swedish Chemicals Policy and Sustainable Development." *Journal of Risk Research* 9:337–60.

Kerret, D., and G.M. Gray. 2007. "What Do We Learn from Emissions Reporting? Analytical Considerations and Comparison of Pollutant Release and Transfer Registers in the United States, Canada, England, and Australia." *Risk Analysis* 27:203–23.

Kinkela, D. 2011. *DDT and the American Century: Global Health, Environmental Politics, and the Pesticide That Changed the World.* Chapel Hill, NC: University of North Carolina Press.

LeChevllier, M.W., and K.-K. Au. 2004. *Water Treatment and Pathogen Control: Process Efficiency in Achieving Safe Drinking Water.* London: IWA Publishing.

Lincer, J.L. 1975. "DDE-Induced Eggshell-Thinning in the American Kestrel: A Comparison of the Field Situation and Laboratory Results." *Journal of Applied Ecology* 12:781–93.

Lipnick, R.L., and D.C.G. Muir. 2000. "History of Persistent, Bioaccumulative, and Toxic Chemicals." In *Persistent, Bioaccumulative, and Toxic Chemicals I: Fate and Exposure,* edited by R.L. Lipnick, J.L.M. Hermens, K. Jones, and D.C.G. Muir. Washington, DC: American Chemical Society.

Löfstedt, R.E. 2003. "Swedish Chemical Regulation: An Overview and Analysis." *Risk Analysis* 23:411–21.

Løkke, S. 2006. "The Precautionary Principle and Chemicals Regulation: Past Achievements and Future Possibilities." *Environmental Science and Pollution Research* 13:342–9.

Maeder, V., B.I. Escher, M. Scheringer, and K. Hungerbühler. 2004. "Toxic Ratio as an Indicator of the Intrinsic Toxicity in the Assessment of Persistent, Bioaccumulative, and Toxic Chemicals." *Environmental Science & Technology* 38:3659–66.

Mansfield, C.A., and B.M. Depro. 2000. "Economic Analysis of Air Pollution Regulations: Chlorine Industry." Research Triangle Park, NC: US Environmental Protection Agency, August. Available at http://www.epa.gov/ttnecas1/regdata/EIAs/Chlorine%20EIA.pdf.

METI (Ministry of the Economy, Trade, and Industry), Government of Japan. 2014. "Policy Information: (1) History of the Law." Available at http://www.meti.go.jp/english/information/data/chemical_substances01.html.

MOE (Ministry of the Environment), Government of Japan. 2002. "Minamata Disease: The History and Measures." Available at http://www.env.go.jp/en/chemi/hs/minamata2002/index.html.

Nixon, R. 1971. "Special Message to the Congress Proposing the 1971 Environmental Program, February 8, 1971." In *Public Papers of the Presidents of the United States: Richard Nixon, 1971,* 125–42. Washington, DC: Government Printing Office, 1972.

NRC (National Research Council). 2009. *Science and Decisions: Advancing Risk Assessment.* Washington, DC: National Academies Press.

OECD (Organization for Economic Cooperation and Development). 2010. *Risk and Regulatory Policy: Improving the Governance of Risk.* Paris: OECD Publishing.

Peakall, D.B., and J.L. Lincer. 1970. "Polychlorinated Biphenyls Another Long-Life Widespread Chemical in the Environment." *BioScience* 20:958–64.

Pilgrim, W., and B. Schroeder. 1997. "Multi-Media Concentrations of Heavy Metals and Major Ions from Urban and Rural Sites in New Brunswick, Canada." *Environmental Monitoring and Assessment* 47:89–108.

Powell, P.P. 1991. "Minamata Disease: A Story of Mercury's Malevolence." *Southern Medical Journal* 84:1352–58.

Raiffa, H. 1970. *Decision Analysis: Introductory Lectures on Choices under Uncertainty.* Reading, MA: Addison-Wesley.

Reich, M.R. 1983. "Environmental Politics and Science: The Case of PBB Contamination in Michigan." *American Journal of Public Health* 73:302–13.

Rosenberg, T. 2004. "What the World Needs Now is DDT." *New York Times*, April 11.

Roueché, B. 1981. *The Medical Detectives*. New York: Truman Talley Books/Plume.

Selin, H. 2010. *Global Governance of Hazardous Chemicals: Challenges of Multilevel Management*. Cambridge, MA: MIT Press.

Shibata, Y., and T. Takasuga. 2007. "Persistent Organic Pollutants Monitoring Activities in Japan." In *Persistent Organic Pollutants in Asia: Sources, Distributions, Transport and Fate*, edited by A. Li, S. Tanabe, G. Jiang, J.P. Giesy, and P.K.S. Lam, 3–30. Boston: Elsevier.

Star, R. 1976. "American and Japanese Controls on Polychlorinated Biphenyls (PCBs)." *Harvard Environmental Law Review* 1:561–7.

Stedeford, T., and M. Banasik. 2009. "International Chemical Control Laws and the Future of Regulatory Testing for Risk Assessment." *Georgetown International Environmental Law Review* 22:619–47.

Sunstein, C.R. 2002. *Risk and Reason: Safety, Law, and the Environment*. New York: Cambridge University Press.

Toda, E. 2009. "Amendment of the Chemical Substance Control Law (CSCL)." Available at http://www.chemical-net.info/pdf/CSCL_Amendment_MrTODA.pdf.

Umeda, G. 1972. "PCB Poisoning in Japan." *Ambio* 1:132–4.

Uyesato, D., M. Weiss, J. Stepanyan, D. Park, K. Yuki, T. Ferris, and L. Bergkamp. 2013. "REACH's Impact in the Rest of the World." In *The European Union REACH Regulation for Chemicals: Law and Practice*, edited by L. Bergkamp, 335–70. New York: Oxford University Press.

van Middelem, C.H. 1966. "Fate and Persistence of Organic Pesticides in the Environment." In *Organic Pesticides in the Environment*, edited by A.A. Rosen and H.F. Kraybill, 228–49. Washington, DC: American Chemistry Society.

van Urk, G., F. Kerkum, and C.J. van Leeuwen. 1993. "Insects and Insecticides in the Lower Rhine." *Water Research* 27:205–13.

van Wijk, D., R. Chénier, T. Henry, M.D. Hernando, and C. Schulte. 2009. "Integrated Approach to PBT and POP Prioritization and Risk Assessment." *Integrated Environmental Assessment and Management* 5:697–711.

WCC (World Chlorine Council). 2002. "The World Chlorine Council and Sustainable Development," April. Available at http://www.worldchlorine .org/wp-content/themes/brickthemewp/pdfs/report.pdf.

Wilson, R., and E.A.C. Crouch. 2001. *Risk-Benefit Analysis*, 2nd ed. Cambridge, MA: Harvard Center for Risk Analysis.

Wilson, T.M., and R.F. Kazmierczak Jr. 2007. "The Public Health and Economic Impacts of Persistent, Bioaccumulative, and Toxic (PBT) Contaminants on U.S. Fisheries." Paper presented at the Annual Meeting of the Southern Agricultural Economics Association, Mobile, AL, February 6–7.

Yokota, F., and K.M. Thompson. 2004. "Value of Information Analysis in Environmental Health Risk Management Decisions: Past, Present, and Future." *Risk Analysis* 24:635–50.

Yoshimura, T. 2003. "Yusho in Japan." *Industrial Health* 41:139–48.

chapter two

PBT determinations
Science and standard procedures

Effective environmental policies require a strong scientific foundation, and so in this chapter, we consider the scientific issues related to the PBT concept, particularly the scientific methods used to assess compounds for persistence, bioaccumulation, and toxicity. As chemical characteristics, persistence, bioaccumulation, and toxicity each have long histories of scientific investigation. The goal of this section is not to provide an extensive review of each of the three properties or a critique of methodological details; such reviews can be found in the literature (e.g., Klečka et al. 2009). Rather, this section highlights important scientific issues that bear directly on PBT determinations. For any chemical, persistence, bioaccumulation, and toxicity are assessed independently using methods appropriate for each characteristic. We therefore address each characteristic individually, noting the issues and concerns specific to that characteristic.

Various approaches can be used to determine the persistence, bioaccumulation, or toxicity of a chemical. The Organization for Economic Cooperation and Development (OECD) has aided in producing standardized testing procedures. The OECD uses expert groups of scientists to develop guidelines for chemical testing. Testing conducted using these guidelines is accepted across OECD nations under the Mutual Acceptance of Data system.* There are currently 34 OECD countries, including the United States, Canada, Japan, and many European countries. For PBTs, the OECD Quantitative Structure–Activity Relationship (QSAR) Toolbox provides means of assessing various characteristics of a chemical, including overall persistence, bioaccumulation, toxicity, and potential for long-range transport. In the following discussions about persistence, bioaccumulation, and toxicity assessments, the OECD guidelines, if applicable, are assumed to be standard procedure.

There is no single or universal test to determine if a chemical is a PBT, nor is there a single or universal test used to determine whether a chemical is persistent, or bioaccumulative, or toxic. The strengths and limitations of any PBT determination may be influenced by the environmental

---

* Results from a chemical safety test conducted in OECD countries shall be accepted by other OECD and adhering countries if the test was carried out according to OECD Test Guidelines and GLP Principles (OECD 2015a).

and health protection goals, the technical methods and tools used for testing, and the integration of testing results within a policy framework.

All measurements have some degree of uncertainty, meaning the true value could be more or less than the measured value, owing to imperfections in measurement tools and techniques. Variability refers to natural differences among measured groups. For example, two populations of a species of fish might differ in their susceptibility to a toxicant because of genetic variability between the populations. Variability and uncertainty can influence any test to evaluate PBT characteristics, and repeated tests often do not produce identical results. Below, we examine the influences of test selection, variability, and uncertainty on the process of PBT determination.

I.   Persistence

Persistence describes the extent to which a substance will remain in the environment without degradation, where degradation refers to transformation of the original chemical structure into a substance that is recognizably different (but not necessarily less hazardous). Mechanisms by which compounds degrade include aerobic and anaerobic biodegradation and abiotic degradation. Biodegradation refers to chemical transformation mediated by the activity of organisms, particularly microorganisms such as fungi and bacteria. Biodegradation can occur in the presence of oxygen (aerobic degradation) or in the absence of oxygen (anaerobic degradation). Abiotic degradation results from chemical and physical reactions that do not require living organisms, such as hydrolysis (reaction with water), photolysis (reaction with light), and oxidation–reduction reactions not catalyzed by organisms. Below, we emphasize biodegradation, although we recognize that abiotic processes can be important in understanding the persistence of organic compounds (OECD 2004b, 2008b).

Because environmental conditions strongly affect the rate and extent to which a compound degrades, it can be challenging to generalize across myriad environmental conditions regarding the persistence of a particular compound. Descriptions of persistence most commonly take the form of estimates of a chemical's half-life—the time required for an initial amount of a chemical to be reduced (degraded) by 50%. A long half-life indicates that a chemical is persistent, whereas a short half-life indicates rapid degradation. Half-lives can be determined for specific media, such as air, water, or soil/sediment. Importantly, half-life values can be readily compared because they are percentages that are not dependent on the initial amount of the chemical used in the test (unless concentration is so large as to inhibit microbiological activity). However, half-life values assume that the concentration of the chemical is not changing, which may or may not be the case in natural settings. Results from laboratory tests for persistence can differ from direct observations in field studies, particularly when transformation rates increase in the field because of increased microbial activity, increased temperatures, or the development of nonextractible residues that can occur when the chemical adsorbs to sediments or soils (Höltge and Kreuzig 2007). Further, there is variation in the approaches to estimating persistence, including the use of estimation methods, laboratory tests, and field monitoring data. The advantages and disadvantages of these different methods are discussed below.

A.   Estimation methods for persistence

Estimation methods model or predict persistence (e.g., half-life of a substance in freshwater) based upon chemical structure and existing data from similar compounds. They can be used to fill in data gaps when no chemical-specific empirical data exist and can be a useful tool in the screening of a large number of substances. Estimation methods are also sometimes used in a confirmatory manner to support or question the accuracy of empirical data.

The US Environmental Protection Agency (US EPA) Office of Pollution Prevention and Toxics, in conjunction with the Syracuse Research Corporation, has developed a screening tool called EPI (Estimation Program Interface) Suite (US EPA 2012a), which calculates various characteristics pertinent to the persistence of a chemical based on its molecular structure. For example, EPI Suite estimates the half-life of substances in various media. Although EPI Suite is not the only estimation model, it is one of the tools used most often, such as in screening exercises involving large numbers of chemicals (Howard and Muir 2010; Rorije et al. 2011; Strempel et al. 2012).

Another estimation program for screening large numbers of chemicals is the OECD QSAR Toolbox (OECD 2015b). The Toolbox primarily uses quantitative structure–activity relationships (QSARs) and was in fact developed in response to regulators' increasing use of QSARs. US EPA (2012b) defines a QSAR as "a mathematical relationship between a quantifiable aspect of chemical structure and a chemical property or reactivity or a well-defined biological activity." Known relationships between chemical structure and biological response can serve as the basis for predicting the effects of chemicals for which empirical data are lacking. The OECD QSAR Toolbox has three main features: identify relevant structural characteristics of a target chemical, identify other chemicals that have the same structural characteristics, and use existing experimental data to fill the data gaps presented by QSARs. The Toolbox can be used to estimate the characteristics of a chemical by using the information from structurally similar chemicals for which information is available (Benfenati 2010).

Methods of estimation offer an efficient means of obtaining data on new chemicals, but they are, by their very nature, estimates with varying degrees of uncertainty. The uncertainty arises mainly from the causal assumptions used in the modeling of persistence and factors such as environmental conditions, rates of biological activity, rates of abiotic reactions, and the applicability of QSARs across compounds. Additionally, persistence is usually represented as a half-life value in a single medium. A single-medium half-life value often cannot adequately describe the complete degradation of a chemical in the environment (Klečka et al. 2009). In this case, a higher-tier screening process such as multimedia fate modeling can, as we discuss below, be used to give an estimate of overall persistence (P_{ov}).

B.   *Laboratory tests for persistence*

Laboratory tests provide empirical measures of persistence and may therefore be preferable to estimation methods. Using standardized protocols for test conditions facilitates comparisons among chemicals, but standard test conditions may not accurately represent the environment(s) to which the chemical will be released. Persistence is rarely a completely intrinsic property, and actual persistence (or degradation) depends upon environmental conditions. Even well-designed laboratory tests are unlikely to capture the full range of environmental conditions that could affect measurements of persistence.

There are four main types of biotic tests routinely used in assessing persistence: ready biodegradability, simulation tests, inherent biodegradability tests, and anaerobic biodegradability screening tests. Abiotic degradation tests are used to examine the potential for hydrolysis and photolysis of a chemical. If there is evidence of rapid hydrolysis or photolysis, then the chemical is unlikely to be persistent in the environment. Generally, OECD test guidelines are the standard approach to these measurements.

Tests for "ready biodegradability" are designed as a pass/fail system, where a positive result is unequivocal (the chemical is not persistent) and a negative result indicates that more in-depth testing should be considered (OECD 2006a). The test does not yield quantitative estimates of persistence but rather indicates when additional testing is needed. There are six possible test methods (not all of which are applicable to each chemical), and each method has a specific pass level that must be reached for a chemical to be considered nonpersistent (OECD 2006a). It is important to select the test that best represents the relevant media that is compatible with the chemical and its potential uses and/or types of releases into the environment.

Ready biodegradability tests are based on a short exposure period (normally 28 days), which allows for the testing to be conducted relatively quickly. Inherent biodegradability tests allow for longer periods of exposure, provide a more in-depth examination of biodegradability, and usually give a more definitive indication of a chemical's persistence (OECD 1981, 1992, 2001a, 2009a). One example is the sewage treatment test (OECD 2001b), which simulates the environment in wastewater treatment plants. The simulation tests can also be completed with respect to soil, sediment, or surface water.* In these tests, samples of soils, sediments, or water are treated with the substance, and the rate of transformation or degradation is measured (OECD 2002). Results from these tests are sensitive to the testing conditions, such as temperature, soil type, and microbial assemblage, which, in some cases, might limit the generality of the results.

All of the biodegradability tests discussed above in Section B are conducted in an aerobic environment. An anaerobic biodegradability test is available to study substances in an oxygen-deficient environment (OECD 2006b). Such environments could include anoxic sediments in lakes, saturated soils, or anaerobic stages of wastewater treatment.

C.   *Monitoring data for persistence*

Environmental monitoring was, in part, responsible for bringing recognition to the fact that some persistent chemicals have the potential for long-range transport (Simonich and Hites 1995; Calamari et al. 2000). The atmospheric transport of organic pollutants to remote environments (e.g., the Arctic) ultimately led to the development of the Stockholm Convention on Persistent Organic Pollutants. Environmental monitoring has therefore played a significant role in the development of policies for PBTs and continues to be used in assessing PBTs.

Environmental monitoring is often conducted over extended periods (i.e., months or years rather than days). Programs that focus on animals or humans measure chemicals in biological samples (Swackhamer et al. 2009). Measurements in humans are most often conducted by analyzing blood (or its components, serum and plasma) or breast milk (Mazdai et al. 2003). This information is regarded as useful for setting priorities for risk assessment/management because it demonstrates some form of actual exposure to the chemicals. When monitoring programs are designed and implemented appropriately, the data, in theory, can support assessments of existing or newly approved chemicals in commerce (e.g., perfluorinated compounds) (Moody et al. 2002). For example, US EPA's *National Study of Chemical Residues in Fish Tissue* includes a robust experimental design to examine the prevalence of 268 PBT contaminants in fish from reservoirs and lakes in the United States (US EPA 2012c).

---

* This is represented in OECD Test Guidelines 307–09 (e.g., OECD 2002).

Monitoring data are limited both in scope and in applicability. In particular, the use of data from the existing literature may result in unintended bias if information regarding content, quality, reliability, and representativeness is lacking. Monitoring data from the literature may have been collected for objectives other than assessing the persistence of a potential PBT, in which case the sampling procedure and the distribution of samples in time and space might be inappropriate for use in a PBT determination. For this reason, detailed methodological information on all monitoring phases, from sample collection to analytical procedures, must be available to determine if the monitoring data are suitable for use in assessing the persistence of a potential PBT.

Pseudo-persistence occurs when a chemical that degrades relatively rapidly, compared to chemicals identified as persistent using the various tests described above, is continuously introduced to an environmental matrix (e.g., air and water) such that the magnitude, frequency, and duration of exposure make the chemical *appear* persistent. Such scenarios are likely relevant for human exposure to indoor air and are particularly important for understanding the exposure(s) to aquatic life in streams with flows dominated by effluent discharges from wastewater treatment plants (Brooks, Riley, and Taylor 2006). In these effluent-dominated streams, effective exposure duration increases when introduction rates from effluent discharges exceed a chemical's rate of degradation (Ankley et al. 2007).

Monitoring data, in general, and well-developed environmental specimen banks, in particular, are useful in determining the accumulation and removal of a chemical over an extended period in the environment and in living organisms (Simonich and Hites 1995; Calamari et al. 2000). However, using monitoring data as a screening tool presents challenges, as discussed previously. Consequently, monitoring data are rarely used alone and should generally be used in conjunction with laboratory tests and estimation methods when making a PBT determination.

D.   *Improvements in methods for assessing persistence*

PBT assessment protocols should be adaptable to improvements and innovations in testing techniques and approaches. In the case of persistence, an example of such an improvement involves determining overall persistence (P_{ov}) in addition to persistence in single media. This concept considers the transfer among different environmental media (multimedia partitioning), recognizing that transport from one medium to another should not be regarded as degradation or disappearance from the environment.

The concept of P_{ov} was developed in the early 1980s (Mackay and Paterson 1982), but its relevance for persistent organic pollutant (POP) and PBT assessment was fully developed only in the last decade (Scheringer

2002). P_{ov} provides an estimate of the time necessary for disappearance of a substance from the overall environment (including all media into which it partitions), the mass of the substance that remains in the environment at a given time after emissions have ceased, and the "turnover time" of the chemical in a flow-through system. Because chemicals can move among media in the environment, knowing persistence in different media is necessary for a comprehensive understanding of potential exposure routes and overall degradation rates.

A distinct advantage of P_{ov} is the ability to compare chemicals on a common scale (Scheringer et al. 2009). Single-medium half-lives are directly comparable only if the chemicals have been tested in a common medium under similar conditions. Even then, the extent to which each chemical will partition into the tested medium could vary in a natural environment. In general, single-medium measurements can yield variable half-lives that are of limited environmental relevance (Webster, Mackay, and Wania 1998). For priority-setting purposes, P_{ov} offers a means to compare and rank chemicals on a common scale and provides a superior estimate of the true persistence of chemicals in the environment.

Whereas single-medium half-lives can be measured through laboratory and field experiments, P_{ov} is much more difficult (if not impossible) to measure experimentally and cannot be estimated from environmental concentration data alone. In general, models of P_{ov} require knowledge or assumptions about partition coefficients, transfer rates among environmental media, and degradation rates in each medium. Often, this information is lacking or incomplete, necessitating the use of multimedia models such as those developed and proposed by OECD (2004a).

The challenges in estimating persistence suggest that caution is warranted in the use of half-life values, and that, as they become available, advances in predicting persistence should be incorporated into PBT assessments. Similarly, incorporating more robust site-specific estimates, including aspects of pseudo-persistence, could reduce uncertainties in persistence determinations associated with differential geographic use, exposure scenarios, and human and ecological risks.

II.   Bioaccumulation

Bioaccumulation occurs when a chemical accumulates in the tissues of an organism; it is the net effect of a set of processes: uptake of a chemical into an organism and loss of the chemical from the organism. Uptake can occur through ingestion, inhalation, or dermal contact. Loss can occur through metabolic breakdown of the chemical or by excretion. A chemical that accumulates in the tissues of an organism has the potential to be transferred from prey to predators, such that the tissue concentrations of the chemical increase from lower trophic levels (i.e., relative position in the food chain) to higher trophic levels in a food web, a phenomenon known as biomagnification. Food webs are descriptors of the environment and the ecological processes occurring in the environment; they explain the potential course that a chemical may take as it moves through the biological components of the system. Chemicals that bioaccumulate and biomagnify are of particular concern because the dietary pathway provides a route of exposure beyond that accounted for in traditional toxicity testing, which is based on aqueous exposure.

With regard to PBT determinations, several metrics are used to predict the bioaccumulative potential of a chemical. The most commonly used metrics are the octanol–water partition coefficient (K_{ow}), the bioconcentration factor (BCF), and the bioaccumulation factor (BAF), each of which is described below. Most authorities use these values to define what chemicals are, or are not, bioaccumulative. Less commonly used metrics include the octanol–air partition coefficient (K_{oa}), the biomagnification factor (BMF), and the trophic magnification factor (TMF). While all of these metrics provide insight into bioaccumulation potential, they each provide slightly different information and require various amounts of time and resources to obtain.

A common and rapid way to estimate a substance's bioaccumulation potential is to determine its octanol–water partition coefficient, K_{ow}, usually expressed on logarithmic scale (log K_{ow}). K_{ow} is a description of how the chemical will partition between octanol (a medium for hydrophobic compounds) and water (a medium for hydrophilic compounds) when at equilibrium. This represents a predictor of bioaccumulation potential because high values of K_{ow} indicate a propensity for a substance to partition into lipids (i.e., lipid soluble), such as fats of fish rather than water. Chemicals that partition into lipids will tend to accumulate in the tissues of an organism, whereas chemicals that partition into water will be more rapidly excreted. The measurement of K_{ow}, however, is a physical–chemical determination and not a direct test for bioaccumulation. For more detailed information on the relationship between K_{ow} and bioaccumulation, see the studies conducted by the Society of Environmental Toxicology and Chemistry Pellston Workshop (Nichols et al. 2009; Weisbrod et al. 2009).

An octanol–air partition coefficient can also be useful for determining the bioaccumulation potential of a substance; however, use of K_{oa} by regulators is not common owing to the historic focus on aquatic systems. K_{oa} predicts the propensity of a chemical to partition from organic media into the air (or vice versa) and therefore is relevant for air-breathing organisms and to estimate absorption in plants because the primary route of exposure can be through the air (Meylan and Howard 2005; Gobas et al. 2009). When assessing potential PBTs, measuring K_{oa} could be useful in screening for bioaccumulation potential in terrestrial organisms where respiratory loss of a chemical is relevant.

Bioconcentration, bioaccumulation, and biomagnification are similar concepts but have distinct endpoints and provide unique information. In simple terms, BCF, BAF, and BMF each reflect competing rates of chemical uptake and elimination in an organism, but they differ in terms of exposure routes and the exposure media used to reference (normalize) the exposure under defined conditions (e.g., organism mass or lipid content).

Bioconcentration is "the process by which a chemical substance is absorbed by an organism from the ambient environment only through its respiratory and dermal surfaces" (Arnot and Gobas 2006, p. 259). Thus, bioconcentration does not consider diet as a route of exposure to a chemical. BCF is the steady-state ratio of a chemical concentration in an organism to its concentration in the surrounding exposure water (preferably the dissolved, bioavailable chemical concentration in the water). BCF is best measured under experimental conditions that ensure that chemical exposure to an organism is from the water only (OECD 2012). Bioaccumulation, on the other hand, accounts for all possible routes of chemical exposure, including diet. Like BCF, BAF is calculated as the steady-state ratio of a chemical concentration in an organism to its concentration in the surrounding exposure water (again, preferably the dissolved, bioavailable chemical concentration in the water).

Biomagnification reflects the degree of bioaccumulation as a result of exposure through the diet only. A BMF is the steady-state ratio of a chemical concentration in an organism to that in its diet (preferably the lipid normalized concentrations). A BMF greater than 1 indicates an increase in concentration of a chemical as it moves up the food chain (i.e., the concentration in the organism is greater than the concentration in its food). A BMF less than 1 indicates a decrease in chemical concentrations with increasing trophic position. Laboratory experiments or field data can be used to determine a BMF. However, it can be difficult to quantify diet and dietary exposure in field studies. Therefore, laboratory studies are preferable because exposure through the diet is controlled and feeding and growth rates can be measured, which facilitates determination of a steady-state value for BMF. A TMF describes the magnification of a chemical in each trophic level of the food chain (Burkhard et al. 2013).

Modeling of each of the metrics described above requires different types of data. Therefore, the particular data on which a bioaccumulation assessment is based will depend on which metric is chosen as the end point. Estimation methods are frequently used because empirical data for bioaccumulation are limited or unavailable. There are, however, numerous laboratory tests that can be completed to help assess bioaccumulation potential. In addition, a few tests can be completed in the field, but the use of field data often is limited because exposure cannot be controlled.

A. Estimation methods for bioaccumulation

Regulators often use estimation methods (e.g., EPI Suite) for screening purposes because empirical data are lacking and the experimental tests needed for determining bioaccumulation are involved and time-consuming. A first step in this process is often to determine the K_{ow} of the chemical under consideration. K_{ow} can be used to estimate BCF, BAF, or BMF because there are linear relationships between log K_{ow} and log BCF and log BMF for various classes of compounds (Jorgensen, Halling-Sorensen, and Mahler 1997). However, at high values of K_{ow}, the linear relationship tends to break down, possibly because transfer rates among media decline and steady-state conditions are not achieved. In most cases, K_{ow} can be estimated using EPI Suite or other computational tools. Based on the molecular composition and structure of the chemical, EPI Suite uses K_{ow} along with SARs to estimate BCF (Lombardo et al. 2010; US EPA 2012a). All of these approaches are based primarily on hydrophobicity and account poorly for any metabolic biotransformation processes. Many chemicals with very high K_{ow} have low bioavailability or are metabolized and excreted (via breakdown products) from living organisms, reducing potential for bioaccumulation (Costanza et al. 2012). Moreover, these estimation methods are not applicable to ionizable chemicals and non-lipophilic bioaccumulative chemicals, such as perfluorinated compounds. Because of the limitations of the estimation methods, relevant indicators of bioaccumulation for potential PBTs may not be accurately estimated by formulae based on K_{ow}.

BCF can also be estimated using chemical read-across, a technique in which the property of a chemical is estimated based on the properties of previously studied chemicals with similar molecular structures. Read-across is based on the assumption that chemicals with similar molecular structures will behave similarly in the environment, but this is not always the case and independent confirmation may be needed in some situations. More details on chemical read-across approaches can be found elsewhere (Verhaar, van Leeuwen, and Hermens 1992; Enoch et al. 2009; Enoch 2010; Sakuratani et al. 2013; OECD 2015b).

B. Laboratory tests for bioaccumulation

Laboratory tests directly measure bioaccumulation in a variety of species and can yield both BCF and BAF estimates. To measure BAF, tests must often control the environment and intake of the substance. There are bioaccumulation test guidelines for fish (including variations for water-soluble and nonwater-soluble compounds) as well as benthic (OECD 2008c) and terrestrial (OECD 2010) oligochaetes (sediment- or soil-dwelling worms). Consideration should be given to selecting appropriate test organisms because BAF values for a given chemical can vary among species and life stages.

The test for BCF or BAF consists of an uptake phase, in which organisms are exposed to a chemical, and a postexposure (depuration) phase, which allows measurement of the chemical's rate of elimination from the organisms. The organisms are monitored throughout the study (usually 28 days), and testing conditions are held constant, as slight variations in water temperature or chemical concentration could affect results (Lombardo et al. 2010).

Although standard methods for determining BCF and BAF exist, empirical data for bioaccumulation potential are limited for the majority of industrial chemicals (Gobas et al. 2009). These tests, excluding the determination of K_{ow}, require significant time to be planned, completed, and analyzed. They also require control and vigilance to ensure that the variation in concentration or temperature remains small and can thus be resource-intensive and expensive to conduct on a large number of chemicals (Weisbrod et al. 2009). The use of animal testing raises ethical considerations and increases the cost of conducting BCF determinations. As a result, testing on fish to determine a BCF is not routinely done, or even allowed, in some jurisdictions. For these reasons, laboratory data are scarce, and authorities often rely on K_{ow} to make estimates of potential for bioaccumulation, despite the fact that K_{ow} is not a direct measurement of bioaccumulation.

C. Monitoring and field data for bioaccumulation

If field data of reasonable quality exist, they may be an important source of information when assessing bioaccumulation potential. However, it is difficult to use field data in a regulatory context because field data often cannot be directly compared with results from laboratory tests, which are the basis for most screening programs (Selck et al. 2012). Nonetheless, there have been several successful fish monitoring studies (Swackhamer and Hites 1988), including the Great Lakes Fish Monitoring and Surveillance Program, which analyzes fish on an annual basis for legacy bioaccumulative compounds (e.g., dieldrin, chlordane, and PCBs) (US EPA 2012d). The basic collection of data occurs by comparing the concentration of specific compounds in fish or other organisms with the concentration in the surrounding media, ideally with control for the age, size, and/or lipid concentration of the study organisms. A biota-sediment accumulation factor is one example for which an extensive literature is available (Burkhard 2009).

Concentration data are quite useful in describing the distribution of a chemical in the environment and in risk assessment and management. However, such data, either in organisms or the environment, are not usually considered sufficient to determine that a chemical is bio-accumulative. There is no harmonized standard protocol across countries for measuring bioaccumulation using field-collected samples. This causes some difficulty in comparing results across studies (Burkhard et al. 2012). Moreover, field data are usually site or habitat specific and therefore may have limited generality to the wider environment. For any field study, information is needed on trophic relationships, organism movement, and habitat characteristics for the data to be interpreted and used to their full extent. As with persistence, the primary exposure pathways to an organism can be difficult to determine solely through use of field data, but those pathways may need to be understood to make a confident determination regarding bioaccumulation.

D.   *Improvements in bioaccumulation measurements*

Although simple measurements related to bioaccumulation are available, such as K_{ow} and BCF, they provide limited insight into the bioaccumulative potential of chemicals in natural settings. Metrics such as BMFs or TMFs, while more difficult to determine, provide much greater insight into the dynamics of bioaccumulation in natural settings.

Recall that BMF is the steady-state ratio of concentrations of a compound at two consecutive trophic levels at steady state. BMF can be calculated from field data on assumed trophic relations or from laboratory feeding experiments and on the assumption that concentrations are close to their steady-state values. This metric is a clear indicator of biomagnification because it shows the relationship between concentrations of a chemical in an organism and its food.

TMF is the average factor by which the normalized chemical concentration in biota increases per trophic level (Hallanger et al. 2011). TMF has been proposed as the best and most reliable tool for assessing bioaccumulation and biomagnification of chemicals (Weisbrod et al. 2009). At present, though, reliable TMFs must be characterized from field studies on chemicals that have been present in the environment for enough time to allow a reasonably close approach to steady state in natural food webs. Therefore, this approach is retrospective and currently cannot be used as a predictive tool. Assessing TMFs through laboratory experiments can be complex, expensive, and time consuming and, thus, this is not widely done, although doing so may provide useful insights (Solomon, Matthies, and Vighi 2013).

BMF and TMF are normalized by lipid content to account for variation in the fat content among animals (OECD 2012). As a result, BMF or TMF values greater than 1 indicate that a chemical will accumulate in higher trophic levels; i.e., the chemical is accumulating in the lipids of the organisms. This provides a definitive answer to the question of whether or not a particular chemical will bioaccumulate. However, TMF is difficult to incorporate into regulatory programs because of uncertainties about the assumption of steady-state conditions and variation among species, ecosystems, and seasons as well as other ecological factors that affect natural populations, such as migratory patterns or disease.

In recent years, researchers have advanced understanding of biotransformation, metabolism (Arnot, Mackay, and Bonnell 2008), and elimination of chemicals by fish. As noted above, estimation of a BCF has historically employed K_{ow} to predict uptake (Veith, DeFoe, and Bergstedt 1979). BCF depends upon the change in chemical concentration in an organism during uptake as well as the change in tissue concentrations during elimination. Thus, elimination rates, despite their limited study in the past, represent a critical component for defining bioaccumulation. *In vitro* assays are increasingly employed to understand metabolic biotransformation by fish. For example, a liver subcellular fraction approach is most commonly used to estimate intrinsic clearance rates for a variety of chemicals (Han et al. 2009; Johanning et al. 2012; Connors et al. 2013). These assays can identify compounds needing more extensive study. They require less time and financial commitments than full BCF studies do and thus can support more robust screening assessments of potentially bioaccumulative chemicals (Nichols et al. 2007). To be successful, the methods need to be standardized, and there is a need for guidance on how the results can be used in the context of PBT determinations.

Additional work also is needed with regard to the bioaccumulation and toxicity of ionizable chemicals (Valenti et al. 2011). Acquisition of a charge tends to make a molecule less lipophilic (soluble in lipids) and more hydrophilic. In addition, because charged molecules have a more difficult time passing through the hydrophobic layers of biological membranes, the time to reach steady-state concentrations may be much slower than for uncharged chemical species. Approaches historically relying on K_{ow}, which were developed for nonionizable chemicals, are not appropriate for chemicals that ionize across environmentally relevant pH (e.g., weak acids, weak bases) (Nakamura et al. 2008; Brooks, Huggett, and Boxall 2009; Fu, Franco, and Trapp 2009; Rendal, Kusk, and Trapp 2011). If the influence of the surface water pH on ionization and subsequent partitioning of a chemical is not considered, estimations of BCFs can be over-predictive or underpredictive, depending on the properties of a chemical and the site-specific characteristics of the matrix (Escher, Schwarzenbach, and Westall 2000b). This can have significant implications for regulatory programs. For example, it has been estimated that over half of the 140,000 chemicals preregistered under Registration, Evaluation, Authorization, and Restriction of Chemicals (REACH) are ionizable and dissociate at environmentally relevant pH (Escher, Schwarzenbach, and Westall 2000a;

Valenti et al. 2009; Franco et al. 2010). For such cases, development of mechanistic models derived from robust uptake and elimination in fish data appear particularly useful to reduce uncertainties associated with bioaccumulation and toxicity assessments (Erickson et al. 2006a, 2006b).

Because of the paucity of empirical data, bioaccumulation assessments continue to rely on predictive tools. For predictive modeling, further improvements can be made by incorporating some of the advances discussed above (e.g., absorption, distribution, metabolism, and excretion measured in fish) into physiologically based pharmacokinetic models (Warne and van Dam 2008; Nichols et al. 2009). As the major factors influencing the movement of a chemical in and out of an organism are captured in models, confidence in models improves. When data permit, incorporating mechanistic understanding of the bioaccumulation phenomenon through the development and testing of mass balance bioaccumulation models allows for better approximation of bioaccumulation and thus can lead to reduction of uncertainties in safety assessment of chemicals.

III.   Toxicity

No single measurement or test can characterize a chemical's toxicity because of the many ways by which a chemical might cause adverse effects and the variability among species and life stages in susceptibility to that chemical. Ideally, toxicity assessments account for a range of adverse outcomes across species and life stages (e.g., reductions in growth of juvenile fish or increased incidence of tumors in rats). Generally then, each system or assay will include one or more endpoints that form the basis to estimate part of the toxicity potential of a given chemical, but assumptions associated with each assay and its endpoints must be carefully linked to safety assessment goals.

Toxicity is most often discussed in terms of acute and chronic exposure conditions, which allows for distinctions between different lengths of the exposure period (e.g., days versus weeks to months). Exposure can be defined by the magnitude, frequency, and duration with which organisms interact with contaminants present in a form that is available for biological uptake. A substance is acutely toxic if it can cause an adverse effect, often mortality, following a short-term exposure (e.g., hours to a few days) to a concentration higher than that required to elicit sublethal responses. Chronic toxicity in organisms can be characterized by a variety of adverse outcomes; thus, regulatory agencies do not all use the same approach when assessing chronic exposures. In most PBT assessments, the doses causing defined adverse effects (e.g., doses causing 50% of the animals to show an effect) in acute or chronic aquatic toxicity tests are used for making toxicity determinations. Aquatic toxicity refers to the effect that the substance has on aquatic organisms representing different components of freshwater or marine ecosystems. The focus on aquatic tox-icity is a result of historical efforts to address water pollution in the 1960s, which led to development of standard toxicity methods using aquatic organisms. Among the various assays available for toxicity testing, *in vivo* assays (whole organism) are the most commonly used, but *in vitro* assays (e.g., cultures of human or animal cells) are increasingly being employed to screen chemicals and identify specific mechanisms of action that may result in adverse outcomes.

Other forms of toxicity assays are intended to determine whether a chemical can cause carcinogenicity, mutagenicity, reproductive/devel-opmental toxicity, and neurotoxicity. A carcinogenic compound has the capability of inducing cancer, whereas mutagenicity occurs when a chem-ical increases the frequency of a mutation in an organism. Reproductive or developmental toxicity refers to the hazards that a chemical poses for reproductive functions, including the development of an embryo or fetus, or even fertility. Often, carcinogenicity, mutagenicity, and reproductive toxicity will be grouped together and labeled as CMR. Neurotoxicity occurs when the chemical induces adverse effects in functioning of the brain, spinal cord, and/or peripheral nervous system. Much study has also been given to understand whether chemicals interfere with under-lying endocrine functions in wildlife and humans. Here, again, specific assays with various endpoints are employed to identify the toxicological pathways (e.g., estrogenicity) that may result in disruption of an endocrine system (e.g., thyroid function).

There are a few common ways in which potencies for these toxic-ity characteristics are measured. Acute aquatic toxicity assays are often described in terms of median lethal concentration (LC_{50}) or half-maximal effective concentration (EC_{50}). The LC_{50} is the concentration of a chemical that kills half of the tested population in a given amount of time (e.g., 48-hour LC_{50}), and EC_{50} is the concentration that induces a biological response (generally adverse) in half of the exposed population. Chronic toxicity criteria employed by regulatory agencies are often described by no observed effect concentration (NOEC) or lowest observed effect con-centration (LOEC) values.* As its name suggests, the NOEC is the highest chemical exposure concentration in a toxicity bioassay that has no statis-tically significant effects in the exposed population relative to the con-trol population. Conversely, the LOEC is the lowest concentration tested at which there is a statistically significant rate of adverse effect in the exposed population. Both of these measures are influenced by the size of the sample population studied and the statistical conventions chosen

---

* These can also be expressed as no or lowest observed adverse effect levels (NOAEL and LOAEL) when the exposure is via oral dose rather than concentration in the aquatic environment.

to determine a "statistically significant" difference between control and exposed groups (and therefore the size of an effect that could be missed). There is some debate about the usefulness of NOEC and LOEC because they are sensitive to the choice of statistical procedures and are influenced by the selection of experimental treatment levels and the number of animals used in dose–response studies (Bailer and Oris 1999; Warne and van Dam 2008; Landis and Chapman 2011; Jager 2012). Some statistically determined low-level endpoints (e.g., effective concentration at 10% inhibition or mortality [EC_{10}]) are often preferred because these measures better characterize exposure levels associated with defined incidences of toxicity (Hanson and Solomon 2002). EC_{50}, LC_{50}, NOEC, and LOEC values are often used as toxicological benchmark concentrations in hazard and risk assessments. Importantly, LC_{50} and EC_{50} are measures of exposure, including nondietary exposure, that cause toxicity and should not be confused with a toxic dose, which is based on consumption of the toxicant.

A.   Estimation methods for toxicity

Empirical toxicity data for the majority of industrial chemicals are lacking, particularly in the publicly available and peer-reviewed literatures. Thus, when toxicity is unknown for a chemical, models are used to estimate potencies for adverse outcomes. These estimation methods for toxicity are often based on QSARs that provide information for several endpoints.

One of the QSAR-based software tools that US EPA uses for toxicity prediction is the Toxicity Estimation Software Tool. This software is used to estimate acute toxicity, developmental toxicity, and mutagenicity (US EPA 2012e). These estimated potencies can then be used for screening purposes or in an assessment when no chemical-specific empirical data exist.

US EPA has also developed a program referred to as Ecological Structure Activity Relationships, or ECOSAR, which uses SARs to predict adverse effects to several organisms and endpoints for untested substances. This is a stand-alone program but has been integrated into the US EPA EPI Suite program for ease of use (US EPA 2014). ECOSAR also provides estimates of chronic toxicity in fish within the EPI Suite. As noted above, endpoints employed to determine chronic toxicity in fish can be numerous, with varying degrees of sensitivity. This information must be used cautiously when mechanisms and modes of action (MOAs) for an untested chemical are unknown. For example, ECOSAR estimates from US EPA's PBT Profiler (discussed in Chapter 6) may be quite useful in predicting chronic toxicity in fish for a chemical exerting toxicity through a narcosis MOA (associated with relatively high lipophilicity as measured by K_{ow}) but would be misapplied for many other receptor-mediated responses (e.g., an estrogen agonist in fish). There are also some classes of chemicals for which toxicity cannot be estimated using ECOSAR (e.g., metals). The toxicity of many organic chemicals (which comprise the majority of PBTs) can be modeled given the underlying toxicity assumptions. But chemicals with certain characteristics, such as high-molecular-weight chemicals (e.g., polymers), cannot be modeled because they are outside the domain of the toxicity models, meaning the models were not developed to address chemicals with those characteristics. Similarly, care must be taken when modeling the toxicity of ionizable chemicals if the underlying information supporting ECOSAR predictions does not account for influences of pH on bioavailability and exposure.

Given the uncertainty associated with QSAR predictions of aquatic toxicity, the exercise typically represents the first level of a tiered risk assessment program. Higher tiers of assessment are usually based on chemical-specific empirical data following multiple chronic toxicity and model ecosystem studies. A variety of options exist to estimate toxic potencies when chemical-specific empirical data are not available. These include chemical approaches, such as computation, interpolation, or categorization (OECD 2015b), and biological approaches, such as cross-species (Berninger and Brooks 2010; Boxall and Ericson 2012; Connors et al. 2013) and read-across techniques (van Leeuwen, Patlewicz, and Worth 2007; Schaafsma et al. 2009; van Leeuwen et al. 2009). Probabilistic approaches based on distributions of chemical toxicity values for specific chemical classes or MOAs are also available (Dobbins, Brain, and Brooks 2008; Dobbins et al. 2009; Berninger et al. 2011; Williams, Berninger, and Brooks 2011). This rapidly expanding area of study provides promise for streamlining assessments, reducing or clarifying uncertainties, and decreasing animal use in toxicity testing.

B. *Laboratory tests for toxicity*

There are many experimental methods for assessing toxicity, both qualitatively and quantitatively, and much variation in testing conditions and the types of tests used to inform toxicity determinations. The value of any toxicity test result hinges on the testing method selected (often termed *measures of effect* in environmental risk assessment), which must be closely linked to the desired assessment endpoints and the environmental protection goals. For example, if a protection goal for a potential PBT is to avoid reproduction impairment in fish, selection of a mutagenicity test is inappropriate. For PBT determinations of toxicity, standardized experimental designs, which specify details to be followed during a toxicity study, are developed by regulatory agencies and academic scientists (Klimisch, Andreae, and Tillmann 1997). As part of the development of standardized methods for toxicity assessments, studies are performed by governmental organizations to characterize variability and assess reproducibility of *in vivo* or *in vitro* responses across various laboratories. Once these studies are performed and deemed acceptable, they are implemented as part of regulatory testing programs (e.g., the endocrine disruption screening program under the US Food Quality Safety Act). Ideally, these experimental requirements are also endorsed or approved by an international organization (e.g., the OECD, the International Organization for Standardization, and the American Society for Testing and Materials), which can then result in consistency and harmonization of data collection by the various global regulatory programs examining PBTs.

The laboratory tests for toxicity can be grouped into ecotoxicity and human health toxicity. In ecotoxicity tests, such as the Fish Acute Toxicity Test, an organism is exposed to a chemical for a short period (e.g., 96 hours), and an LC_{50} is determined. There is significant variation among agencies and jurisdictions in testing requirements for the determination of toxicity for a potential PBT. In some cases, only aquatic toxicity is specified in the endpoints, although terrestrial data could be helpful in many cases. Some chemicals are not amenable to standard toxicity tests because of poor solubility in water or other characteristics. The OECD (2000) has addressed this issue and provided guidance on conducting aquatic toxicity tests for "difficult to test" substances. For the majority of industrial chemicals, very few species are examined (sometimes as few as three), which can reduce the generality of the results of toxicity assessment.

Both acute toxicity and chronic toxicity are considered when assessing human health. Acute toxicity analyzes the relationship between dose and adverse effects during a short exposure period (OECD 1996). Another type of dose–response study design is the short-term (14, 21, or 28 days) repeated-dose toxicity study. In this testing approach, animals are exposed to repeated doses of the substance, which aids in identifying delayed effects and the specific organs affected by the chemical (OECD 2008a). Chronic toxicity is used to identify the adverse outcomes of repeated and constant exposure to a substance for an extended period; the OECD (2009b) test guideline for chronic toxicity requires a period of 12 months. At the conclusion of these studies, statistical analysis is employed to provide NOEC and LOEC values for the various toxicological endpoints. To a certain extent, a well-designed set of toxicity tests may account for persistent and bioaccumulative chemicals (e.g., a chronic bioassay with repeated exposure may replicate a highly persistent substance).

Toxicity testing requires use of animals and, in some cases, the length of time required for the tests is extensive. However, such long-term toxicity data, which are derived from chronic exposures, are important because these tests focus on actual adverse effects that may arise from the persistent nature of PBTs. One limitation with toxicity assessments, particularly for PBTs, is that the assays normally are conducted in an aqueous environment and therefore provide data only for aquatic organisms. This is a concern for chemicals that are highly water-insoluble and suggests care is needed when selecting toxicity tests for potential PBTs.

Higher-tier assessment methods (e.g., mesocosms, field, and semifield studies) are used successfully in ecotoxicology and provide a powerful tool for improving the ecological realism of toxicity assessments. Mesocosm studies have also been successfully applied for assessing BCF and TMF (Solomon et al. 2009). Drawbacks to using higher-tier assessments for regulatory purposes include the variability of ecological responses and the need for transparency in the evaluation of the results. In particular, a priority for research is the development of statistically based tools capable of quantitatively assessing uncertainties and to improve the transparent use of these approaches (SCHER, SCENIHR, and SCCS 2013).

Much like persistence and bioaccumulation, the science of toxicity assessments continues to advance. For example, US EPA recently proposed the Adverse Outcome Pathway (AOP) concept for use in ecological risk assessments of chemicals (Ankley et al. 2010). An AOP is a conceptual framework that portrays existing knowledge of chemical action and of an adverse outcome at a level of biological organization relevant to risk assessment (e.g., individual level for threatened and endangered species, population level for other organisms). This approach is built logically on previous contributions (Bradbury, Feijtel, and van Leeuwen 2004; Ankley et al. 2007; Krewski et al. 2010) and provides an approach to plausibly link mechanistic toxicology and ecology. The use of AOPs represents one path forward for future chemical safety assessments and OECD is increasingly including AOPs in their assessment programs.

As noted above, toxicity data are simply not available for the majority of industrial chemicals (EDF 1997; Schaafsma et al. 2009; Williams, Panko, and Paustenbach 2009), although numerous national and international programs (e.g., REACH and the US High Production Volume Challenge program) are beginning to address some of the deficiencies. Because testing all chemicals with the full range of available toxicity tests is not feasible, screening approaches are necessary to identify potential hazardous substances and prioritize those compounds for more intensive study. US EPA's innovative computational toxicology programs are focusing on generating high-throughput screening data for many substances. For example, US EPA's Tox21 and ToxCast programs are collectively screening thousands of industrial chemicals with hundreds of *in vitro* toxicity model systems in an effort to identify which structural attributes of chemicals may result in specific AOPs.* This information will eventually be used to inform the development of computational models for prediction of chemical properties resulting in various types of toxicity.

---

* Tox21 is available online at http://epa.gov/ncct/Tox21/. ToxCast is available online at http://epa.gov/ncct/toxcast/.

These approaches, when coupled with efforts to design less hazardous chemicals—an important principle of green chemistry—can leverage lessons learned from developing and testing safer pharmaceuticals to advance sustainable molecular design of less toxic industrial chemicals (Anastas and Warner 2000; Voutchkova et al. 2011; Voutchkova-Kostal et al. 2012).

IV.   PBT determinations

A.   Standard regulatory approaches

A variety of estimation, laboratory, and field methods are used to measure the persistence, bioaccumulation, and toxicity properties of chemicals, and it is useful to explore how regulatory authorities can use this information in actual PBT determinations. The different methods can be considered at hierarchical levels of integration.

The most definitive information comes from laboratory and field data if the tests adequately reflect real-world environmental conditions, the testing parameters are described, and the data are of sufficient quality. Whenever possible, these data should be considered in any PBT determination. Of course, the availability of field data is dependent on a chemical already being present in the environment, so these data are usually not available for new chemicals. Laboratory data are the most definitive source of information at the design phase. When chemical-specific field and laboratory data are unavailable or of insufficient quality, or when a rapid assessment is necessary, estimation methods such as QSARs are often used.

In a PBT determination, each property is generally assessed relative to a cutoff value. For example, bioaccumulation is often represented by a BCF. If the BCF is greater than a predetermined cutoff value, the substance is considered bioaccumulative. Although there is general consensus about the metrics, the cutoff values vary among organizations and users. Similarly, it is clear from our review that specific toxicity tests and the ways these types of toxicity data are used for PBT determinations are not consistent globally. Table 2.1 presents the cutoff criteria adopted by different organizations, countries, and regions. It also includes information on long-range transport, which is a criterion in POP determinations.

One interesting aspect of the PBT determination process is the origin and evolution of the various cutoff values. Although the concepts of persistence, bioaccumulation, and toxicity have a sound scientific basis, the cutoff values were not selected solely on the basis of empirical science. Each component is a continuous variable, and therefore, no obvious cutoff values distinguish a PBT from a non-PBT. The selection of the cutoff values, therefore, has been a hybrid science-policy decision. Originally, there was concern about a small group of compounds known to be persistent, bioaccumulative,

Table 2.1 Criteria for identifying PBTs and POPs used by various organizations[a]

Definition	Persistence	Bioaccumulation	Toxicity	Potential for long-range transport
Convention on Long-range Transboundary Air Pollution: POP criteria[b]	Half-life in water > 2 months, sediment > 6 months, soil > 6 months, **or** evidence that chemical is otherwise sufficiently persistent	BCF > 5000, BAF > 5000, **or** Log K_{ow} > 5	Potential to adversely affect human health or the environment	Vapor pressure < 1000 Pa and half-life in air > 2 days, **or** detection from monitoring data in remote area
Stockholm Convention: POP criteria[c]	Half-life in water > 2 months, sediment > 6 months, soil > 6 months, **or** evidence that chemical is otherwise sufficiently persistent	BCF > 5000, BAF > 5000, Log K_{ow} > 5, monitoring data indicating bioaccumulation potential in biota, **or** evidence of other reasons for concern	Evidence of adverse effect on human health or the environment or toxicity characteristics indicating potential damage to human health or the environment	Measured levels far from source, **or** monitoring data in remote area, **or** multimedia modeling evidence and half-life in air > 2 days
Oslo-Paris Convention: PBT criteria[d]	Not readily biodegradable or half-life in water ≥ 50 days	BCF ≥ 500 **or** Log K_{ow} ≥ 4	Acute aquatic toxicity L(E)C$_{50}$ ≤ 1 mg/L **or** long-term NOEC ≤ 0.1 mg/L, **or** mammalian toxicity: CMR or chronic toxicity	Not applicable

(Continued)

Table 2.1 (Continued) Criteria for identifying PBTs and POPs used by various organizations[a]

Definition	Persistence	Bioaccumulation	Toxicity	Potential for long-range transport
EU, REACH: PBT criteria[e]	Half-life in marine water > 60 days, fresh water > 40 days, marine sediment > 180 days, fresh water sediment > 120 days, **or** soil > 120 days	BCF > 2000	Chronic NOEC < 0.01 mg/L **or** CMR **or** endocrine disrupting effects	Not applicable
EU, REACH: vPvB criteria[e]	Half-life in marine or fresh water > 60 days, marine or fresh water sediment > 180 days, **or** soil > 180 days	BCF > 5000	Not applicable	Not applicable
USA, Toxics Release Inventory: PBT criteria[f]	Half-life in soil, sediment, or water ≥ 2 months, **or** in air ≥ 2 days	BCF ≥ 1000 **or** BAF ≥ 1000		Not applicable
USA, Toxic Substances Control Act (TSCA), New Chemicals Program: PBT subject to control action, pending testing[g]	Half-life in water, soil, or sediment > 2 months	BCF ≥ 1000 **or** BAF ≥ 1000	Develop toxicity data where necessary based upon various factors, including concerns for persistence, bioaccumulation, other physicochemical factors, and toxicity based on existing data	Not applicable

(Continued)

Table 2.1 (Continued) Criteria for identifying PBTs and POPs used by various organizations[a]

Definition	Persistence	Bioaccumulation	Toxicity	Potential for long-range transport
USA, TSCA, New Chemicals Program: PBT subject to ban, pending testing[g]	Half-life in water, soil, or sediment > 6 months	BCF ≥ 5000		Not applicable
Japan, Chemical Substances Control Law: PBT criteria[h]	Not readily biodegradable	BCF > 5000, if BCF is between 1000 and 5000, sources other than BCF are consulted	Poses a risk to human health or the environment if ingested or exposed continuously	Not applicable
Canadian Environmental Protection Act: PBiT criteria[i]	Half-life in air ≥ 2 days, subject to transport to remote areas, water ≥ 182 days, sediment ≥ 365 days, or soil ≥ 182 days	BAF ≥ 5000, BCF ≥ 5000, or Log K_{ow} ≥ 5	Not set in policy or regulations; set as 1 mg/L (acute aquatic) or 0.1 mg/L (chronic aquatic) as part of categorization of substances on Domestic Substances List	Not applicable as a separate criterion but included as part of persistence criterion

(Continued)

Table 2.1 (Continued) Criteria for identifying PBTs and POPs used by various organizations[a]

Definition	Persistence	Bioaccumulation	Toxicity	Potential for long-range transport
California, Green Chemistry Initiative[i]	Half-life in marine, fresh, estuary water > 40–60 days, soil or sediment > 2 months, air > 2 days, **or** identification as persistent by an authoritative organization	Evidence of adverse effect on human health or the environment	BMF or TMF > 1 for aquatic or terrestrial organisms, BAF or BCF > 1000 Log K_{ow} > 4, or Log K_{oa} > 5, **or** identification as bioaccumulative by an authoritative organization	

[a] This table is a modified version of that found in van Wijk et al. (2009) and European Commission (2009).

[b] UNECE, Executive Body Decision 1998/2 on Information to be Submitted and the Procedure for Adding Substances to Annexes I, II, or III to the Protocol on Persistent Organic Pollutants.

[c] Stockholm Convention, Annex D.

[d] Convention for the Protection of the Marine Environment of the North-East Atlantic, Sept. 22, 1992, 2354 U.N.T.S. 67. See OSPAR Commission (2005).

[e] REACH, Annex XIII.

[f] US EPA. Persistent Bioaccumulative Toxic (PBT) Chemicals; Final Rule, 64 Fed. Reg. 58,666, 58,669 (Oct. 29, 1999).

[g] US EPA. Category for Persistent, Bioaccumulative, and Toxic New Chemical Substances, 64 Fed. Reg. 60,194, 60,202 (Nov. 4, 1999).

[h] Kashinho (Chemical Substances Control Law), Law No. 117 of 1973. See also Nakai et al. (2006) and Hashizume et al. (2013).

[i] Persistence and Bioaccumulation Regulations (Canadian Environmental Protection Act 1999), SOR/2000-107. Under the Canadian Environmental Protection Act 1999, "toxic" has a distinct regulatory meaning; the iT abbreviation—standing for *inherent* toxicity—is used in the Canadian context to refer to toxic effects. See Chapter 5 for additional details.

[j] California Code of Regulations, title 22, division 4.5, Chapter 54, Art. 5, §69405 (2014).

and toxic (e.g., the "dirty dozen") (Meek and Armstrong 2007; Lipnick and Muir 2000). There were some patterns among these compounds in terms of persistence, bioaccumulation, and toxicity, and these patterns suggested cutoff values that would capture those known harmful compounds.

From this starting point, modification of the cutoff values can capture more or fewer chemicals within a regulatory framework. Cutoff values in REACH, for example, are designed to include more chemicals in the PBT designation than those of the Stockholm Convention. European lawmakers considered a half-life of six months or lower for persistence and a BCF of 5000 or lower for bioaccumulation—the values used in the Stockholm Convention. However, they decided to set REACH's cutoff values for persistence and bioaccumulation at a two-month half-life and a BCF of 2000, respectively, to capture more chemicals. This decision was informed by science but was ultimately a hybrid science-policy decision.

When examining lists of chemicals identified as PBTs, it is important to recognize whether partial PBTs are included. In addition to PBTs, regulatory programs or legislative action may also target chemicals that are only PT or BT, but not persistent, bioaccumulative, and toxic. Approaches that include partial PBTs create much larger lists of chemicals. Indeed, Environment Canada identified thousands of potential PTs and BTs in the categorization effort, but far fewer met all three PBT criteria. Given the enormous amount of resources that it would take to conduct full risk assessments on the thousands of potential PTs and BTs, the Canadian government opted to prioritize its efforts first on screening-level assessments for PBTs and other priority chemicals (e.g., potentially harmful substances with likely high exposure to humans).

Alternatively, regulations may target chemicals that are persistent and bioaccumulative, but not definitively toxic, as in the very persistent, very bioaccumulative (vPvB) designation in REACH. Here, the goal is to identify those chemicals that are so long-lived and prone to bioaccumulation that they warrant special consideration. For example, toxicity is generally assumed for vPvB substances because these chemicals are often insoluble in water or have other properties that preclude standard aquatic toxicity testing. Thus, it must be emphasized that the PBT determination process—and the number of chemicals classified as PBTs and/or partial PBTs—is influenced by both the objectives and resource constraints of the sponsoring organizations.

The original set of chemicals of concern with respect to persistence and bioaccumulation consisted mainly of chlorinated pesticides and PCBs. Yet, there are a number of other groups of chemical substances that may be determined to be PBTs. Among the most controversial groups are metals (and metalloids, which will be included here by the term "metals"), and debate continues as to whether metals should be addressed when conducting PBT determinations. Metals are, practically speaking, infinitely persistent, and therefore, measures of persistence are not applicable. In addition, some of the standardized tests are not applicable for metals, precluding the completion of normal assessment procedures. As a result, there is no consistent treatment of metals among regulatory organizations. In some cases, organo-metal compounds are assessed for PBT properties, but the metal alone is not. In other cases, metals are considered as PBTs, but as a separate category to make the distinction known. To some extent, the same concern applies to silicon-based substances and new physical forms of carbon-based chemicals, such as nanotubes.

B.   Uncertainty in PBT determinations

All empirical and modeling results are associated with some degree of measurement error or other uncertainties. Uncertainty or error associated with specific measurements, such as a half-life, K_{ow}, or an LC_{50}, can be readily quantified. This type of uncertainty is not unique to PBT determinations, and there are statistical methods for attempting to address this type of uncertainty.

While recognized as important, uncertainty remains a challenging factor to incorporate into PBT determinations. For example, there is no practical difference between a BCF of 4000 and 6000 (both values indicate a high potential to partition into organisms) except that one does not exceed a cutoff of 5000 and the other does. The establishment of cutoff values for persistence, bioaccumulation, and toxicity establishes "bright lines" for regulatory purposes but can also create a false sense of confidence in the outcome of a PBT determination. It has been recognized that bright lines and rigid application of cutoff values may not always result in correctly identifying PBTs and non-PBTs. Socolow (1976) has observed that uncertain values can become "golden numbers." He writes:

> Environmental discourse likewise manifests a powerful dependence on numbers. A number that may once have been an effusion of a tentative model evolves into an immutable constraint. Apparently, the need to have precision in the rules of the game is so desperate that the administrators seize on numbers (in fact, get legislators to write them into laws) and then carefully forget where they come from. (Socolow 1976, p. 7)

Furthermore, reliance solely on established persistence, bioaccumulation, and toxicity criteria may result in the omission of relevant information (e.g., biomonitoring data) during a PBT determination.

One way to include a broader range of information and to avoid the potential pitfalls of a strictly bright-line approach, is to use a weight of evidence (WOE) approach. The growing importance of WOE reasoning in PBT determinations is addressed Chapter 3.

A difficult challenge in PBT assessment is the uncertainty associated with projection from modeling and laboratory studies to real-world environments. Implicit in the concept of PBT is the notion that if a chemical is designated a PBT, the designation applies to all environments under all conditions. However, environmental contexts can strongly influence how quickly a chemical degrades or bioaccumulates. Toxicity is also, to some extent, contextual because not all species, populations, or individuals will be equally susceptible to the effects of a chemical, thus the importance of incorporating US EPA's AOP approach. No laboratory or modeling result truly simulates all aspects of a real environment, and no chemical can feasibly be tested for PBT properties under all environments and conditions in which it might be released. Thus, treating persistence, bioaccumulation, and toxicity as intrinsic properties of a chemical is understood as a practical simplification of a more complex reality that is impossible to replicate in standard testing.

References

Anastas, P.T., and J.C. Warner. 2000. *Green Chemistry: Theory and Practice*. New York: Oxford University Press.

Ankley, G.T., B.W. Brooks, D.B. Huggett, and J.P. Sumpter. 2007. "Repeating History: Pharmaceuticals in the Environment." *Environmental Science and Technology* 41:8211–7.

Ankley, G.T., R.S. Bennett, R.J. Erickson, D.J. Hoff, M.W. Hornung, R.D. Johnson, D.R. Mount et al. 2010. "Adverse Outcome Pathways: A Conceptual Framework to Support Ecotoxicology Research and Risk Assessment." *Environmental Toxicology and Chemistry* 29:730–41.

Arnot, J.A., and F.A.P.C. Gobas. 2006. "A Review of Bioconcentration Factor (BCF) and Bioaccumulation Factor (BAF) Assessments for Organic Chemicals in Aquatic Organisms." *Environmental Reviews* 14:257–97.

Arnot, J.A., D. Mackay, and M. Bonnell. 2008. "Estimating Metabolic Biotransformation Rates in Fish from Laboratory Data." *Environmental Toxicology and Chemistry* 27:341–51.

Bailer, A.J., and J.T. Oris. 1999. "What is a NOEC? Non-Monotonic Concentration-Response Patterns Want to Know." *SETAC News*, March 1999:22–4.

Benfenati, E. 2010. "QSAR Models for Regulatory Purposes: Experiences and Perspectives." In *Practical Aspects of Computational Chemistry*, edited by J. Leszczynski and M.K. Shukla 183–200. New York: Springer.

Berninger, J.P., and B.W. Brooks. 2010. "Leveraging Mammalian Pharmaceutical Toxicology and Pharmacology Data to Predict Chronic Fish Responses to Pharmaceuticals." *Toxicology Letters* 193:69–78.

Berninger, J.P., B. Du, K.A. Connors, S.A. Eytcheson, M.A. Kolkmeier, K.N. Prosser, T.W. Valenti, C.K. Chambliss, and B.W. Brooks. 2011. "Effects of the Antihistamine Diphenhydramine on Selected Aquatic Organisms." *Environmental Toxicology and Chemistry* 30:2065–72.

Boxall, A.B.A., and J.F. Ericson. 2012. "Environmental Fate of Human Pharmaceuticals." In *Human Pharmaceuticals in the Environment*, edited by B.W. Brooks and D.B. Huggett, 63–83. New York: Springer.

Bradbury, S.P., T.C.J. Feijtel, and C.J. van Leeuwen. 2004. "Meeting the Scientific Needs of Ecological Risk Assessment in a Regulatory Context." *Environmental Science & Technology* 38:463A–70A.

Brooks, B.W., T.M. Riley, and R.D. Taylor. 2006. "Water Quality of Effluent-Dominated Ecosystems: Ecotoxicological, Hydrological, and Management Considerations." *Hydrobiologia* 556:365–79.

Brooks, B.W., D.B. Huggett, and A.B.A. Boxall. 2009. "Pharmaceuticals and Personal Care Products: Research Needs for the next Decade." *Environmental Toxicology and Chemistry* 28:2469–72.

Burkhard, L. 2009. "Estimation of Biota Sediment Accumulation Factor (BSAF) From Paired Observations of Chemical Concentrations in Biota and Sediment." U.S. Environmental Protection Agency, Ecological Risk Assessment Support Center, Cincinnati, OH. EPA/600/R-06/047.

Burkhard, L.P., J.A. Arnot, M.R. Embry, K.J. Farley, R.A. Hoke, M. Kitano, H.A. Leslie et al. 2012. "Comparing Laboratory and Field Measured Bioaccumulation Endpoints." *Integrated Environmental Assessment and Management* 8:17–31.

Burkhard, L.P., K. Borgå, D.E. Powell, P. Leonards, D.C.G. Muir, T.F. Parkerton, and K.B. Woodburn. 2013. "Improving the Quality and Scientific Understanding of Trophic Magnification Factors (TMFs)." *Environmental Science & Technology* 47:1186–7.

Calamari, D., K. Jones, K. Kannan, A. Lecloux, M. Olsson, M. Thurman, and P. Zannetti. 2000. "Monitoring as an Indicator of Persistence and Long-Range Transport." In *Evaluation of Persistence and Long-Range Transport of Organic Chemicals in the Environment*, edited by G. Klečka, B. Boethling, J. Franklin, L. Grady, D. Graham, P.H. Howard, K. Kannan, R.J. Larson, D. Mackay, D. Muir, and D. van de Meent, 205–39. Pensacola, FL: Society of Environmental Toxicology and Chemistry.

Connors, K.A., B. Du, P.N. Fitzsimmons, A.D. Hoffman, C.K. Chambliss, J.W. Nichols, and B.W. Brooks. 2013. "Comparative Pharmaceutical Metabolism by Rainbow Trout (Oncorhynchus Mykiss) Liver S9 Fractions." *Environmental Toxicology and Chemistry* 32:1810–8.

Costanza, J., D.G. Lynch, R.S. Boethling, and J.A. Arnot. 2012. "Use of the Bioaccumulation Factor to Screen Chemicals for Bioaccumulation Potential." *Environmental Toxicology and Chemistry* 31:2261–8.

Dobbins, L.L., R.A. Brain, and B.W. Brooks. 2008. "Comparison of the Sensitivities of Common in vitro and in vivo Assays of Estrogenic Activity: Application of Chemical Toxicity Distributions." *Environmental Toxicology and Chemistry* 27:2608–16.

Dobbins, L.L., S. Usenko, R.A. Brain, and B.W. Brooks. 2009. "Probabilistic Ecological Hazard Assessment of Parabens Using Daphnia Magna and Pimephales Promelas." *Environmental Toxicology and Chemistry* 28:2744–53.

Enoch, S.J. 2010. "Chemical Category Formation and Read-across for the Prediction of Toxicity." In *Recent Advances in QSAR Studies: Methods and Applications*, edited by T. Puzyn, J. Leszczynski, and M.T.D. Cronin, 209–19. Springer, New York.

Enoch, S.J., M.T.D. Cronin, J.C. Madden, and M. Hewitt. 2009. "Formation of Structural Categories to Allow for Read-Across for Teratogenicity." *QSAR & Combinatorial Science* 28:696–708.

EDF (Environmental Defense Fund). 1997. "Toxic Ignorance: The Continuing Absence of Basic Health Testing for Top-Selling Chemicals in the United States." Available at http://www.edf.org/sites/default/files/243_toxicignorance _0.pdf.

Erickson, R.J., J.M. McKim, G.J. Lien, A.D. Hoffman, and S.L. Batterman. 2006a. "Uptake and Elimination of Ionizable Organic Chemicals at Fish Gills: I. Model Formulation, Parameterization, and Behavior." *Environmental Toxicology and Chemistry* 25:1512–21.

Erickson, R.J., J.M. McKim, G.J. Lien, A.D. Hoffman, and S.L. Batterman. 2006b. "Uptake and Elimination of Ionizable Organic Chemicals at Fish Gills: II. Observed and Predicted Effects of pH, Alkalinity, and Chemical Properties." *Environmental Toxicology and Chemistry* 25:1522–32.

Escher, B.I., R.P. Schwarzenbach, and J.C. Westall. 2000a. "Evaluation of Liposome-Water Partitioning of Organic Acids and Bases. 1. Development of a Sorption Model." *Environmental Science & Technology* 34:3954–61.

Escher, B.I., R.P. Schwarzenbach, and J.C. Westall. 2000b. "Evaluation of Liposome-Water Partitioning of Organic Acids and Bases. 2. Comparison of Experimental Determination Methods." *Environmental Science & Technology* 34: 3962–8.

European Commission. 2009. "Proposal to Consider the Harmonisation of the Criteria for Classification and Labelling of Persistent, Bioaccumulative and Toxic (PBT) and Very Persistent and Very Bioaccumulative (vPvB) Substances." United Nations Economic Commission for Europe Sub-Committee of Experts on the Globally Harmonized System of Classification and Labelling of Chemicals. Informal Documents 18th Sess. UN/SCEGHS /18/INF.4. Available at http://www.unece.org/fileadmin/DAM/trans/doc /2009/ac10c4/UN-SCEGHS-18-inf04e.pdf.

Franco, A., A. Ferranti, C. Davidsen, and S. Trapp. 2010. "An Unexpected Challenge: Ionizable Compounds in the REACH Chemical Space." *International Journal of Life Cycle Assessment* 15:321–5.

Fu, W., A. Franco, and S. Trapp. 2009. "Methods for Estimating the Bioconcentration Factor of Ionizable Organic Chemicals." *Environmental Toxicology and Chemistry* 28:1372–9.

Gobas, F.A.P.C., W. de Wolf, L.P. Burkhard, E. Verbruggen, and K. Plotzke. 2009. "Revisiting Bioaccumulation Criteria for POPs and PBT Assessments." *Integrated Environmental Assessment and Management* 5:624–37.

Hallanger, I.G., N.A. Warner, A. Ruus, A. Evenset, G. Christensen, D. Herzke, G.W. Gabrielsen, and K. Borgå. 2011. "Seasonality in Contaminant Accumulation in Arctic Marine Pelagic Food Webs Using Trophic Magnification Factor as a Measure of Bioaccumulation." *Environmental Toxicology and Chemistry* 30:1026–35.

Han, X., D.L. Nabb, C.H. Yang, S.I. Snajdr, and R.T. Mingoia. 2009. "Liver Microsomes and S9 from Rainbow Trout (Oncorhynchus Mykiss): Comparison of Basal-Level Enzyme Activities with Rat and Determination of Xenobiotic Intrinsic Clearance in Support of Bioaccumulation Assessment." *Environmental Toxicology and Chemistry* 28:481–8.

Hanson, M.L., and K.R. Solomon. 2002. "New Technique for Estimating Thresholds of Toxicity in Ecological Risk Assessment." *Environmental Science & Technology* 36:3257–64.

Hashizume, N., Y. Inoue, H. Murakami, H. Ozaki, A. Tanabe, Y. Suzuki, T. Yoshida, E. Kikushima, and T. Tsuji. 2013. "Resampling the Bioconcentration Factors Data from Japan's Chemical Substances Control Law Database to Simulate and Evaluate the Bioconcentration Factors Derived from Minimized Aqueous Exposure Tests." *Environmental Toxicology and Chemistry* 32:406–9.

Höltge, S., and R. Kreuzig. 2007. "Laboratory Testing of Sulfamethoxazole and Its Metabolite Acetyl-Sulfamethoxazole in Soil." *CLEAN–Soil, Air, Water* 35:104–10.

Howard, P.H., and D.C.G. Muir. 2010. "Identifying New Persistent and Bio-accumulative Organics among Chemicals in Commerce." *Environmental Science & Technology* 44:2277–85.

Jager, T. 2012. "Bad Habits Die Hard: The NOEC's Persistence Reflects Poorly on Ecotoxicology." *Environmental Toxicology and Chemistry* 31(2):228–9.

Johanning, K., G. Hancock, B.I. Escher, A. Adekola, M.J. Bernhard, C.E. Cowan-Ellsberry, J. Domoradzki et al. 2012. "Assessment of Metabolic Stability Using the Rainbow Trout (Oncorhynchus Mykiss) Liver S9 Fraction." In *Current Protocols in Toxicology*, 14.10.1–28. New York: John Wiley & Sons.

Jorgensen, S.E., B. Halling-Sorensen, and H. Mahler. 1997. *Handbook of Estimation Methods in Ecotoxicology and Environmental Chemistry*. Boca Raton, FL: CRC Press.

Klečka, G.M., D.C.G. Muir, P. Dohmen, S.J. Eisenreich, F.A.P.C. Gobas, K.C. Jones, D. Mackay, J.V. Tarazona, and D. van Wijk. 2009. "Introduction to Special Series: Science-Based Guidance and Framework for the Evaluation and Identification of PBTs and POPs." *Integrated Environmental Assessment and Management* 5:535–8.

Klimisch, H.-J., M. Andreae, and U. Tillmann. 1997. "A Systematic Approach for Evaluating the Quality of Experimental Toxicological and Ecotoxicological Data." *Regulatory Toxicology and Pharmacology* 25:1–5.

Krewski, D., D. Acosta Jr., M. Andersen, H. Anderson, J.C. Bailar III, K. Boekelheide, R. Brent et al. 2010. "Toxicity Testing in the 21st Century: A Vision and a Strategy." *Journal of Toxicology and Environmental Health*, 13:51–138.

Landis, W.G., and P.M. Chapman. 2011. "Well Past Time to Stop Using NOEL/LOELs." *Integrated Environmental Assessment and Management* 7:vi–viii.

Lipnick, R.L., and D.C.G. Muir. 2000. "History of Persistent, Bioaccumulative, and Toxic Chemicals." In *Persistent, Bioaccumulative, and Toxic Chemicals I: Fate and Exposure*, edited by R.L. Lipnick, J.L.M. Hermens, K. Jones, and D.C.G. Muir, 1–12. Washington, DC: American Chemical Society.

Lombardo, A., A. Roncaglioni, E. Boriani, C. Milan, and E. Benfenati. 2010. "Assessment and Validation of the CAESAR Predictive Model for Bioconcentration Factor (BCF) in Fish." *Chemistry Central Journal* 4(Suppl 1):S1–11.

Mackay, E., and S. Paterson. 1982. "Fugacity Revisited." *Environmental Science & Technology* 16:654A–60A.

Mazdai, A., N.G. Dodder, M.P. Abernathy, R.A. Hites, and R.M. Bigsby. 2003. "Polybrominated Diphenyl Ethers in Maternal and Fetal Blood Samples." *Environmental Health Perspectives* 111:1249–52.

Meek, M.E., and V.C. Armstrong. 2007. "The Assessment and Management of Industrial Chemicals in Canada." In *Risk Assessment of Chemicals: An Introduction*, edited by C.J. van Leeuwen, and T.G. Vermeire, 591–621. Dordrecht, The Netherlands: Springer.

Meylan, W.M., and P.H. Howard. 2005. "Estimating Octanol–Air Partition Coefficients with Octanol–Water Partition Coefficients and Henry's Law Constants." *Chemosphere* 61:640–4.

Moody, C.A., J.W. Martin, W.C. Kwan, D.C.G. Muir, and S.A. Mabury. 2002. "Monitoring Perfluorinated Surfactants in Biota and Surface Water Samples Following an Accidental Release of Fire-Fighting Foam into Etobicoke Creek." *Environmental Science & Technology* 36:545–51.

Nakai, S., K. Takano, and S. Saito. 2006. "Bioconcentration Prediction under the Amended Chemical Substances Control Law of Japan." *Trans. Sumitomo Kagaku* 2006-I. Sumitomo Chemical Co., Environmental Health Science Laboratory. Available at http://www.sumitomo-chem.co.jp/english/rd/report/theses /docs/20060106_vpv.pdf.

Nakamura, Y., H. Yamamoto, J. Sekizawa, T. Kondo, N. Hirai, and N. Tatarazako. 2008. "The Effects of pH on Fluoxetine in Japanese Medaka (*Oryzias latipes*): Acute Toxicity in Fish Larvae and Bioaccumulation in Juvenile Fish." *Chemosphere* 70:865–73.

Nichols, J., S. Erhardt, S. Dyer, M. James, M. Moore, K. Plotzke, H. Segner, I. Schultz et al. 2007. "Use of in vitro Absorption, Distribution, Metabolism, and Excretion (ADME) Data in Bioaccumulation Assessments for Fish." *Human and Ecological Risk Assessment* 13:1164–91.

Nichols, J., M. Bonnell, S.D. Dimitrov, B.I. Escher, X. Han, and N.I. Kramer. 2009. "Bioaccumulation Assessment Using Predictive Approaches." *Integrated Environmental Assessment and Management* 5:577–97.

OECD (Organization for Economic Cooperation and Development). 1981. "Test No. 302A: Inherent Biodegradability: Modified SCAS Test." In *OECD Guidelines for the Testing of Chemicals, Section 3: Degradation and Accumulation*. Paris: OECD.

OECD (Organization for Economic Cooperation and Development). 1992. "Test No. 302B: Inherent Biodegradability: Zahn-Wellens/EVPA Test." In *OECD Guidelines for the Testing of Chemicals, Section 3: Degradation and Accumulation*. Paris: OECD.

OECD (Organization for Economic Cooperation and Development). 1996. "Test No. 422. Combined Repeated Dose Toxicity Study with the Reproduction/ Developmental Toxicity Screening Test." In *OECD Guidelines for the Testing of Chemicals, Section 4: Health Effects*. Paris: OECD.

OECD (Organization for Economic Cooperation and Development). 2000. "Guidance Document on Aquatic Toxicity Testing of Difficult Substances and Mixtures." OECD Series on Testing and Assessment, No. 23. ENV/JM/ MONO(2000)6. Paris: OECD.

OECD (Organization for Economic Cooperation and Development). 2001a. "Test No. 303: Simulation Test—Aerobic Sewage Treatment—A: Activated Sludge Units; B: Biofilms." In *OECD Guidelines for the Testing of Chemicals, Section 3: Degradation and Accumulation*. Paris: OECD.

OECD (Organization for Economic Cooperation and Development). 2001b. "Proposal for a New Guideline 302D: Inherent Biodegradability—CONCAWE Test." In *OECD Guidelines for the Testing of Chemicals, Draft Document*. Paris: OECD.

OECD (Organization for Economic Cooperation and Development). 2002. "Test No. 307: Aerobic and Anaerobic Transformation in Soil." In *OECD Guidelines for the Testing of Chemicals, Section 3: Degradation and Accumulation*. Paris: OECD.

OECD (Organization for Economic Cooperation and Development). 2004a. "Guidance Document on the Use of Multimedia Models for Estimating Overall Environmental Persistence and Long-Range Transport." OECD Series on Testing and Assessment, No. 45. ENV/JM/MONO(2004)5. Paris: OECD.

OECD (Organization for Economic Cooperation and Development). 2004b. "Test No. 111: Hydrolysis as a Function of pH. Paris: Organisation for Economic Co-operation and Development." In *OECD Guidelines for the Testing of Chemicals, Section 1: Physical-Chemical Properties*. Paris: OECD.

OECD (Organization for Economic Cooperation and Development). 2006a. "Revised Introduction to the OECD Guidelines for Testing of Chemicals, Section 3." In *OECD Guidelines for the Testing of Chemicals, Section 3: Degradation and Accumulation*. Paris: OECD.

OECD (Organization for Economic Cooperation and Development). 2006b. "Test No. 311: Anaerobic Biodegradability of Organic Compounds in Digested Sludge: By Measurement of Gas Production." In *OECD Guidelines for the Testing of Chemicals, Section 3: Degradation and Accumulation*. Paris: OECD.

OECD (Organization for Economic Cooperation and Development). 2008a. "Test No. 407: Repeated Dose 28-Day Oral Toxicity Study in Rodents." In *OECD Guidelines for the Testing of Chemicals, Section 4: Health Effects*. Paris: OECD.

OECD (Organization for Economic Cooperation and Development). 2008b. "Test No. 316: Phototransformation of Chemicals in Water—Direct Photolysis." In *OECD Guidelines for the Testing of Chemicals, Section 3: Degradation and Accumulation*. Paris: OECD.

OECD (Organization for Economic Cooperation and Development). 2008c. "Test No. 315: Bioaccumulation in Sediment-Dwelling Benthic Oligochaetes." In *OECD Guidelines for the Testing of Chemicals, Section 3: Degradation and Accumulation*. Paris: OECD.

OECD (Organization for Economic Cooperation and Development). 2009a. "Test No. 452: Chronic Toxicity Studies." In *OECD Guidelines for the Testing of Chemicals, Section 4: Health Effects*. Paris: OECD.

OECD (Organization for Economic Cooperation and Development). 2009b. "Test No. 302C: Inherent Biodegradability: Modified MITI Test (II)." In *OECD Guidelines for the Testing of Chemicals, Section 3: Degradation and Accumulation*. Paris: OECD.

OECD (Organization for Economic Cooperation and Development). 2010. "Test No. 317: Bioaccumulation in Terrestrial Oligochaetes." In *OECD Guidelines for the Testing of Chemicals, Section 3: Degradation and Accumulation*. Paris: OECD.

OECD (Organization for Economic Cooperation and Development). 2012. "Test No. 305: Bioaccumulation in Fish: Aqueous and Dietary Exposure." In *OECD Guidelines for the Testing of Chemicals, Section 3: Degradation and Accumulation*. Paris: OECD.

OECD (Organization for Economic Cooperation and Development). 2015a. "Mutual Acceptance of Data (MAD)." Accessed January 7. Available at http://www.oecd.org/env/ehs/testing/mutualacceptanceofdatamad.htm.

OECD (Organization for Economic Cooperation and Development). 2015b. "The OECD QSAR Toolbox." Accessed January 7. Available at http://www.oecd .org/env/ehs/risk-assessment/theoecdqsartoolbox.htm.

OSPAR Commission. 2005. "Cut-Off Values for the Selection Criteria of the OSPAR Dynamic Selection and Prioritisation Mechanism for Hazardous Substances." *Meeting of the OSPAR Commission*, Malahide, Ireland, June 27– July 1. 05/21/1-E, Annex 7.

Rendal, C., K.O. Kusk, and S. Trapp. 2011. "Optimal Choice of pH for Toxicity and Bioaccumulation Studies of Ionizing Organic Chemicals." *Environmental Toxicology and Chemistry* 30:2395–406.

Rorije, E., E.M.J. Verbruggen, A. Hollander, T.P. Traas, and M.P.M. Janssen. 2011. "Identifying Potential POP and PBT Substances: Development of a New Persistence/Bioaccumulation-Score." National Institute for Public Health and the Environment. RIVM Report 601356001/2011. Bilthoven, The Netherlands. Available at http://www.rivm.nl/bibliotheek/rapporten /601356001.pdf.

Sakuratani, Y., H.Q. Zhang, S. Nishikawa, K. Yamazaki, T. Yamada, J. Yamada, K. Gerova, G. Chankov, O. Mekenyan, and M. Hayashi. 2013. "Hazard Evaluation Support System (HESS) for Predicting Repeated Dose Toxicity Using Toxicological Categories." *SAR and QSAR in Environmental Research* 24:351–63.

Schaafsma, G., E.D. Kroese, E.L.J.P. Tielemans, J.J.M. van de Sandt, and C.J. van Leeuwen. 2009. "REACH, Non-Testing Approaches and the Urgent Need for a Change in Mind Set." *Regulatory Toxicology and Pharmacology* 53:70–80.

SCHER (Scientific Committee on Health and Environmental Risks), SCENIHR (Scientific Committee on Emerging and Newly Identified Health Risks), and SCCS (Scientific Committee on Consumer Safety). 2013. *Addressing the New Challenges for Risk Assessment*. Brussels: European Commission. Available at http://ec.europa.eu/health/scientific_committees/consumer_safety/docs /sccs_o_131.pdf.

Scheringer, M. 2002. *Persistence and Spatial Range of Environmental Chemicals: New Ethical and Scientific Concepts for Risk Assessment*. New York: John Wiley & Sons.

Scheringer, M., K.C. Jones, M. Matthies, S.L. Simonich, and D. van de Meent. 2009. "Multimedia Partitioning, Overall Persistence, and Long-Range Transport Potential in the Context of POPs and PBT Chemical Assessments." *Integrated Environmental Assessment and Management* 5:557–76.

Selck, H., K. Drouillard, K. Eisenreich, A.A. Koelmans, A. Palmqvist, A. Ruus, D. Salvito et al. 2012. "Explaining Differences between Bioaccumulation Measurements in Laboratory and Field Data through Use of a Probabilistic Modeling Approach." *Integrated Environmental Assessment and Management* 8:42–63.

Simonich, S.L., and R.A. Hites. 1995. "Global Distribution of Persistent Organo-chlorine Compounds." *Science* 269:1851–54.

Socolow, R.H. 1976. "Failures of Discourse: Obstacles to the Integration of Environmental Values into Natural Resource Policy: A Reading of the Controversy Surrounding the Proposed Tocks Island Dam on the Delaware River." In *When Values Conflict: Essays on Environmental Analysis, Discourse, and Decision*, edited by L.H. Tribe, C.S. Schelling, and J. Voss, 1–33. Cambridge, MA: Ballinger.

Solomon, K.R., P. Dohmen, A. Fairbrother, M. Marchand, and L. McCarty. 2009. "Use of (Eco) Toxicity Data as Screening Criteria for the Identification and Classification of PBT/POP Compounds." *Integrated Environmental Assessment and Management* 5:680–96.

Solomon, K.R., M. Matthies, and M. Vighi. 2013. "Assessment of PBTs in the European Union: A Critical Assessment of the Proposed Evaluation Scheme with Reference to Plant Protection Products." *Environmental Sciences Europe* 25:1–17.

Strempel, S., M. Scheringer, C.A. Ng, and K. Hungerbühler. 2012. "Screening for PBT Chemicals among the 'Existing' and 'New' Chemicals of the EU." *Environmental Science & Technology* 46:5680–7.

Swackhamer, D.L., and R.A. Hites. 1988. "Occurrence and Bioaccumulation of Organochlorine Compounds in Fishes from Siskiwit Lake, Isle Royale, Lake Superior." *Environmental Science & Technology* 22:543–8.

Swackhamer, D.L., L.L. Needham, D.E. Powell, and D.C.G. Muir. 2009. "Use of Measurement Data in Evaluating Exposure of Humans and Wildlife to POPs/PBTs." *Integrated Environmental Assessment and Management* 5:638–61.

US EPA. 2012a. "Estimating Physical/Chemical and Environmental Fate Properties with EPI Suite." In *Sustainable Futures/P2 Framework Manual.* Office of Chemical Safety and Pollution Prevention. EPA-748-B12-001. Washington, D.C. Available at http://www.epa.gov/oppt/sf/pubs/epi-pchem-fate.pdf.

US EPA. 2012b. "Glossary of Terms: Methods of Toxicity Testing and Risk Assessment." Last modified May 9. Available at http://www.epa.gov/opp00001/science/comptox-glossary.html#q.

US EPA. 2012c. "National Lake Fish Tissue Study." Last modified June 15. Available at http://water.epa.gov/scitech/swguidance/fishstudies/lakefishtissue_index.cfm.

US EPA. 2012d. "Great Lakes Fish Monitoring and Surveillance Program." Last modified October 17. Available at http://www.epa.gov/glnpo/monitoring/fish/index.html.

US EPA. 2012e. "User's Guide for T.E.S.T. (version 4.1) (Toxicity Estimation Software Tool): A Program to Estimate Toxicity from Molecular Structure." Available at http://www.epa.gov/nrmrl/std/qsar/TEST-user-guide-v41.pdf.

US EPA. 2014. "Ecological Structure Activity Relationships (ECOSAR)." Last modified January 13. Available at http://www.epa.gov/oppt/newchems/tools/21ecosar.htm.

Valenti, T.W., P. Perez-Hurtado, C.K. Chambliss, and B.W. Brooks. 2009. "Aquatic Toxicity of Sertraline to Pimephales Promelas at Environmentally Relevant Surface Water pH." *Environmental Toxicology and Chemistry* 28:2685–94.

Valenti, T.W., J.M. Taylor, J.A. Back, R.S. King, and B.W. Brooks. 2011. "Influence of Drought and Total Phosphorus on Diel pH in Wadeable Streams: Implications for Ecological Risk Assessment of Ionizable Contaminants." *Integrated Environmental Assessment and Management* 7:636–47.

van Leeuwen, C.J., G.Y. Patlewicz, and A.P. Worth. 2007. "Intelligent Testing Strategies." In *Risk Assessment of Chemicals: An Introduction,* edited by C.J. van Leeuwen, and T.G. Vermeire, 467–509. Dordrecht, The Netherlands: Springer.

van Leeuwen, C.J., T.W. Schultz, T. Henry, B. Diderich, and G.D. Veith. 2009. "Using Chemical Categories to Fill Data Gaps in Hazard Assessment." *SAR and QSAR in Environmental Research* 20:207–20.

van Wijk, D., R. Chénier, T. Henry, M.D. Hernando, and C. Schulte. 2009. "Integrated Approach to PBT and POP Prioritization and Risk Assessment." *Integrated Environmental Assessment and Management* 5:697–711.

Veith, G.D., D.L. DeFoe, and B.V. Bergstedt. 1979. "Measuring and Estimating the Bioconcentration Factor of Chemicals in Fish." *Journal of the Fisheries Board of Canada* 36:1040–8.

Verhaar, H.J.M., C.J. van Leeuwen, and J.L.M. Hermens. 1992. "Classifying Environmental Pollutants." *Chemosphere* 25:471–91.

Voutchkova, A.M., J. Kostal, J.B. Steinfeld, J.W. Emerson, B.W. Brooks, P.T. Anastas, and J.B. Zimmerman. 2011. "Towards Rational Molecular Design: Derivation of Property Guidelines for Reduced Acute Aquatic Toxicity." *Green Chemistry* 13:2373–9.

Voutchkova-Kostal, A.M., J. Kostal, K.A. Connors, B.W. Brooks, P.T. Anastas, and J.B. Zimmerman. 2012. "Towards Rational Molecular Design for Reduced Chronic Aquatic Toxicity." *Green Chemistry* 14:1001–8.

Warne, M.S.J., and R. van Dam. 2008. "NOEC and LOEC Data Should No Longer Be Generated or Used." *Australasian Journal of Ecotoxicology* 14:1–5.

Webster, E., D. Mackay, and F. Wania. 1998. "Evaluating Environmental Persistence." *Environmental Toxicology and Chemistry* 17:2148–58.

Weisbrod, A.V., K.B. Woodburn, A.A. Koelmans, T.F. Parkerton, A.E. McElroy, and K. Borgå. 2009. "Evaluation of Bioaccumulation Using in vivo Laboratory and Field Studies." *Integrated Environmental Assessment and Management* 5:598–623.

Williams, E.S., J. Panko, and D.J. Paustenbach. 2009. "The European Union's REACH Regulation: A Review of Its History and Requirements." *Critical Reviews in Toxicology* 39:553–75.

Williams, E.S., J.P. Berninger, and B.W. Brooks. 2011. "Application of Chemical Toxicity Distributions to Ecotoxicology Data Requirements under REACH." *Environmental Toxicology and Chemistry* 30:1943–54.

chapter three

PBT determinations

Weight of evidence approaches and the number of PBTs in commerce

I. Weight of evidence

Weight of evidence (WOE) is a widely used concept, particularly in the field of risk assessment. However, a recent study of how the concept is used in risk assessment found no clear and consistent definition (Weed 2005). The author, in fact, found that risk assessors use the concept in several different ways. Proponents of WOE assessment urge organizations to clearly define what is meant by WOE and how it is going to be used by scientists and decision makers.

In 1986, for example, the US Environmental Protection Agency (US EPA 2014) introduced WOE reasoning into its guidelines for carcinogen risk assessment. In this application, WOE referred to the need to consider positive and negative evidence of carcinogenicity in both animals and humans and to evaluate bodies of biological evidence for alternative modes of action. Ultimately, US EPA applied WOE judgment in the classification of substances with respect to potential for carcinogenic hazard (known, probable, possible, unlikely, or inadequate evidence to assess).

WOE reasoning is also used in the study of ecosystem impairments (Burton, Chapman, and Smith 2002). An ecological risk assessment may examine constructs called lines of evidence. This involves examining individual characteristics or processes (the "lines of evidence") and then combining them in a comprehensive manner to make an overall conclusion that is informed by all of the lines of evidence (Hull and Swanson 2006). A researcher examining sediment quality, for example, might use a multivariate approach to gather information about chemical concentrations in the sediment, aquatic life living in potentially impaired water, and ambient toxicity tests for benthic invertebrates. This common WOE approach, known as the sediment quality triad, allows a researcher to draw conclusions from multiple lines of evidence. To many practicing ecologists, toxicologists, and environmental scientists, WOE is state-of-the-art.

In the case of PBT determinations, WOE does not have a precise definition, but it is typically understood as a more flexible, integrative approach that differs from a bright-line approach based solely on numerical cutoff values from specified tests for persistence, bioaccumulation, and toxicity. Moreover, as applied in the case of PBTs, WOE refers to use of all available data—including evaluations of data reliability and relevance—before making a PBT determination. This is particularly important for PBT determinations because data on persistence, bioaccumulation, and toxicity are often absent or incomplete for many proposed PBTs. Thus, WOE means that more information, and different kinds of information and judgments, may be considered than would be the case under the traditional bright-line approach. Moreover, reliance on nonstandard data (e.g., biomonitoring studies) increases the chances of conflicting results and the consequent need for WOE. Below are examples of situations that may warrant a WOE approach to PBT determinations:

- Laboratory tests and models produce different results for persistence, bioaccumulation, or toxicity.
- A substance has low toxicity in standard laboratory tests, but epidemiological evidence or ecological studies suggest adverse effects in particular settings or for specific species.
- Biomonitoring data (or other field data) suggest persistence or bioaccumulation even though estimation methods or laboratory tests show low potential for persistence or bioaccumulation.
- Data from studies of varying strength and relevance need to be assessed and incorporated into the PBT determination.

The WOE terminology has been used in PBT and persistent organic pollutant (POP) deliberations. In discussions during the development of the Convention on Long-Range Atmospheric Transport, there was a dispute about whether there should be strict numerical criteria or whether WOE judgment should be permitted. The United States and Canada advocated strict numerical criteria because this would add transparency and predictability to the determinations, whereas several northern European countries preferred WOE assessment because it would allow more flexibility over time to account for advances in science (Selin 2010). A compromise was reached that is embodied in the Stockholm Convention. Instead of including specific tests and cutoff values in the treaty's language, the criteria and cutoff values are included in an Annex, which allows for additional tests and expert judgment to be used for identifying POPs. Interestingly, the specific term *weight of evidence* was not used in the convention, but government scientists reportedly use the term when discussing whether a particular chemical should be added to the list of POPs.

A. WOE and Registration, Evaluation, Authorization, and Restriction of Chemicals Annex XIII

European Union (EU) authorities have recently highlighted the relevance of WOE reasoning to PBT assessment (Solomon, Matthies, and Vighi 2013). In 2011, the language in Annex XIII of Registration, Evaluation, Authorization, and Restriction of Chemicals (REACH) was modified for PBT determinations to include WOE. One driver for this change was the concern that reliance solely on cutoff values would not capture all of the worrisome chemicals that authorities wanted to target. Likewise, using only cutoff values (rather than all relevant information) might misclassify some chemicals as PBTs. Instead of relaxing the cutoff values again, to capture more chemicals of concern, application of WOE approaches was seen as a way to provide more leeway in the determination of persistence, bioaccumulation, and toxicity. Further, advances in biomonitoring are generating relevant data about persistence and bioaccumulation that were not readily usable when relying only on cutoff values from specific tests. Thus, WOE was also seen as providing the option to consider advances in science such as new biomonitoring data. Finally, before REACH, some Member States were already using WOE to make PBT determinations (at least informally) and, thus, the modification of Annex XIII was seen as a codification of prior practice.

Annex XIII states that when a WOE approach is warranted, it should be conducted using expert judgment to compare all relevant and applicable data to the criteria for persistence, bioaccumulation, and toxicity. Annex XIII also indicates that, in a WOE determination, "the quality and consistency of the data shall be given appropriate weight." The purpose of including WOE in Annex XIII is not to circumvent the established criteria for persistence, bioaccumulation, and toxicity, nor to allow inclusion of low-quality data. Unfortunately, it is not yet clear exactly how WOE will be used in PBT determinations under Annex XIII. In fact, the European Chemicals Agency's (ECHA's) guidance on making PBT determinations now embraces WOE judgment without providing indications of how WOE should be employed (ECHA 2014; Müller 2014).

B. Developing guidance for a WOE approach to PBTs

In the PBT context, WOE assessment without any formal guidance is fraught with problems related to subjectivity, credibility, and lack of predictability. The absence of guidance also is a step backward from the perspective of international transparency, learning, and harmonization. It is therefore encouraging that the PBT Expert Group under REACH is considering guidance on how WOE should inform PBT determinations. We suggest that, in the long run, a WOE approach will prove to be preferable to strict numerical cutoffs, assuming that formal guidance is developed, applied, and modified over time to ensure incorporation of scientific advances, technical integrity, transparency, and a measure of predictability.

WOE assessment can be applied to the separate determinations of persistence, bioaccumulation, and toxicity for a given chemical. An example is the use of the fugacity ratio approach for determining bioaccumulation (Burkhard et al. 2012). By standardizing field and laboratory estimates of bioaccumulation, the fugacity ratio allows for a robust WOE determination of the potential for a chemical to bioaccumulate. As another example, the biomagnification factor (BMF) for a substance may support a positive determination concerning bioaccumulation, even though the traditional criteria and cutoff value for bioaccumulation refer to the bioconcentration factor (BCF), rather than BMF (see Chapter 2 for BCF and BMF descriptions). In addition, the BCF cutoff value for bioaccumulation is defined for aquatic animals, but evidence for bioaccumulation could arise from studies of terrestrial animals. WOE allows for consideration of nonstandard data, such as BMF, data from terrestrial animals, or other field data (Rauert et al. 2014).

Guidance is also needed as to when monitoring (including biomonitoring) information is relevant or determinative with respect to persistence or bioaccumulative determinations. It is not clear whether—or under what circumstances—negative monitoring data should override positive indications of persistence from standard tests or whether it is relevant only in borderline cases. Nor is it clear when positive monitoring results should support affirmative persistence or bioaccumulative determinations if results for the traditional criteria are negative.

In the case of a persistence determination, the mere presence of a chemical in the environment (or even in human blood samples) does not necessarily establish that the chemical is persistent. Such measurements could be an indication of continuous releases of the chemical—releases that might be controlled adequately through conventional risk management. On the other hand, if longitudinal information on releases and environmental presence is available, it may be feasible to make a convincing persistence determination with biomonitoring information alone.

For bioaccumulation determinations, field data on biomagnification are already being used, under the discretion of WOE, to make positive findings, even when substances do not meet the required BCF according to traditional tests and, conversely, to exclude substances that exceed the established cutoff values. But guidance is needed on how biomagnification data could be used to support a very bioaccumulative (vB) determination, as cutoff values for a vB determination based on biomagnification data have not been established.

In all cases, attention must be given to the quality and applicability of the data used in a WOE approach. All data have some degree of uncertainty, and this should be recognized. Field data (e.g., biomonitoring results or environmental occurrence) gleaned from the scientific literature may be site specific and thus of limited generality to support a PBT determination. Not all data in the scientific literature will be of equal quality and applicability to a particular PBT determination. Additionally, field studies often vary in both field sampling protocols and analytical methods, which can reduce the comparability of data from different studies. Guidance specific to PBTs is needed regarding the inclusion or exclusion of data from the PBT determination process.

One might assume that WOE assessment will result in a greater number of positive PBT determinations than if regulators relied only on cutoff values. In the long run, however, the result could go the other way, depending on how the science evolves and is applied. For example, biomonitoring data could lead to down-rating the priority of some substances for which measured values for persistence and bioaccumulation are only slightly above the cutoff values. In the vast majority of cases, however, we expect WOE and bright-line approaches to produce the same PBT determinations.

Although it is not tailored to PBTs, ECHA (2010) has developed a practical guidance document on how to use WOE in judgments on substance properties. The guidance document defines how WOE judgment may be used in decision-making processes under REACH. There are four contributing factors: reliability, adequacy, relevance, and quantity. The guide describes how each of these factors should be considered in WOE decision making. The European Commission's Scientific Committee on Emerging and Newly Identified Health Risks (SCENIHR) has also addressed WOE with respect to human health and risk assessment. SCENIHR (2012) provides detailed guidance on using scientific literature in a WOE approach. As with ECHA's guidance, SCENIHR recommends assessing the relevance and quality of each published study. SCENIHR then recommends combining these characteristics into a ranking of low, medium, or high utility, with utility defined as the combination of relevance and quality. Along with utility, each line of evidence is also ranked as high, medium, or low for consistency. High consistency indicates that most of the studies produce the same result/conclusion, while low consistency indicates little agreement among studies. In the end, each study used in the determination is ranked for utility, and each line of evidence is ranked for consistency. This approach provides rigor to WOE when relying on published literature. It also provides a mechanism for transparency when using WOE to reach a determination about specific chemicals.

One notable strength of the SCENIHR approach is the explicit recognition that different types of studies can be used to address the same question or topic. In this regard, SCENIHR (2012) provides specific criteria for the evaluation of each of the various types of studies used in human health risk assessments, such as epidemiologic studies, human biomarker studies, animal studies, and others. For PBT determinations, this level of guidance could be quite helpful when relying on results or data from the scientific literature. For example, the question of whether a particular substance is bioaccumulative can be addressed with laboratory tests, field experiments, field monitoring data, or models (e.g., Quantitative Structure–Activity Relationships [QSARs]). Additionally, laboratory tests can vary in terms of specific methods and test conditions. Should one type of study be given greater weight or considered more reliable than others? An advantage of WOE is the option to include nonstandard types of information. Thus, we suggest that studies be assessed on their internal strengths and weaknesses, as recommended by SCENIHR, rather than assuming that certain study types are inherently better than others.

In reviewing the use of WOE in risk assessment and, more specifically, for PBT determinations, we identified four fundamental concepts that we believe should be reported with each PBT determination based on WOE. These include the following:

1. Completeness of the assessment. All studies and data sets used to reach a PBT determination should be reported, and an effort should be made to identify and consider all potentially relevant information. Furthermore, the databases, search engines, and query terms used to search the literature should be reported. Any secondary analyses, such as a meta-analysis, should be described in sufficient detail to allow the analysis to be independently verified.
2. Ranking of the evidence. It is likely that some studies or lines of evidence will influence a final PBT determination more so than others will. This should be reported through the ranking or categorizing of each study or line of evidence, for example, as described by SCENIHR (2012).
3. Strength of the determination. The final PBT determination should be described in terms of its overall strength. In some cases, the weight of the evidence may be overwhelming, while in other cases, it may be moderate or even weak but still sufficient to reach a determination. At minimum, a narrative should be developed that explains the strength of the determination.
4. Description of uncertainty. All WOE-based PBT determinations are associated with some degree of uncertainty. Uncertainty can be assessed informally in a descriptive manner or formally using statistical approaches. In either case, uncertainty should be recognized as it is essentially the inverse of the strength of determination described above. Consideration of uncertainly also reveals the important data gaps that should be filled to improve the strength of the determination.

Applying these concepts, along with others identified in scientific literature (e.g., Rhomberg, Bailey, and Goodman 2010; Borgert et al. 2011), to PBT determinations would be fruitful and contribute to reducing uncertainty and vagueness in the application of WOE to PBTs. Even if these concepts are fully considered, other issues may emerge in a WOE-based determination that require description. Examples might include the extent to which expert judgment was relied upon or the details of how conflicting lines of evidence were reconciled.

C. *WOE and the problem of "partial" PBTs*

A potentially problematic use of WOE concerns chemicals for which measured values of persistence, bioaccumulation, and/or toxicity are very close to the cutoff values. If a chemical is clearly persistent and toxic but has a measured BCF just below the cutoff value, some may argue that it should nonetheless be considered a PBT. After all, persistence, bioaccumulation, and toxicity are continuous constructs, and there is no empirical basis for preferring any particular cutoff value.

The flaw in this reasoning is that it defeats the purpose of having cutoff values at all. The purpose of cutoff values—and of the PBT construct as a policy term of art—is to help set priorities for regulators by limiting the number of substances under consideration to a manageable quantity. Instead of relaxing cutoff values on a case-by-case basis, the disciplined approach would be to relax the cutoff values in a general way, once the existing set of PBTs has been properly assessed and managed and once it is apparent that more resources are available to assess chemicals of lower priority.

For substances that do not satisfy the cutoff values for PBTs (i.e., partial PBTs), REACH provides alternative paths for appropriate regulation. One possibility is to classify a substance as of "equivalent concern" to a PBT or CMR, which means the substance could be listed as a substance of very high concern (SVHC), thereby triggering the stringent authorization process (see Chapter 4). In this case, an affirmative PBT determination is not necessary if "equivalent concern" is demonstrated. Before ECHA and the EU move in this direction, it would be wise to provide guidance that ensures some scientific rigor, credibility, and predictability in determinations of "equivalent concern." In general, one would expect that substances that possess all three properties (persistence, bioaccumulation, and toxicity) would be of greater concern than partial PBT substances that have only one or two of these properties.

Perhaps a better approach to addressing unacceptable risks from partial PBTs (and even some substances of equivalent concern) may be regulation under REACH's restrictions authority, without making a PBT finding or an SVHC listing under REACH. If there are unacceptable risks, the commission may initiate necessary protective measures, typically on a use-by-use basis and with support from risk assessment and socioeconomic analysis. We discuss the use of restrictions authority more fully in Chapter 4.

D. Summary and case studies

In summary, WOE is a tricky concept for PBT determinations—sound in theory but challenging in real-world application, especially without any guidelines. Specifically, the change in Annex XIII of REACH has increased the quest for clarity about how WOE will be used for PBT determinations. On the one hand, PBT assessment and determination should be predictable so that industry is able to make investment decisions in production and innovation. On the other hand, WOE reasoning allows for improvements in science to be considered, and therefore, the regulatory determinations can be modified based on up-to-date data.

Some apprehension about WOE assessment arises from the potential for subjectivity and lack of transparency. If WOE is used in a systematic and rigorous way, the potential drawbacks can be minimized without forgoing the benefits of scientific discretion. This might be achieved through the development of guidelines through an open process of consultation, especially if common guidelines are established and implemented consistently across jurisdictions. To be successful, this process should include programs for training policy makers and regulatory scientists on the proper application of WOE reasoning.

To highlight some of the issues that we have raised about WOE assessment of PBTs, we present three case studies. The first case study demonstrates how WOE can identify PBTs that would not have been classified as such under strict adherence to cutoff values. The second case study illustrates how WOE assessment can reverse a PBT determination where the cutoff values have been met and how authorities in different jurisdictions can arrive at inconsistent conclusions on PBT determinations by assessing the weight of the evidence. Although this book is focused primarily on industrial chemicals, the final case study examines a pesticide (lindane) identified as a POP under the Stockholm Convention. We broaden our scope to include a pesticide because the determination of lindane as a POP illustrates a situation in which authorities had to reconcile conflicting lines of evidence.

1. Case study: Henicosafluoroundecanoic acid

The henicosafluoroundecanoic acid case illustrates how WOE reasoning can be used to classify a chemical as a PBT (in this instance very persistent, very bioaccumulative [vPvB]) when the relevant test results do not satisfy the cutoff criteria. As we noted above, the discretion embodied in

WOE allows relevant information to be considered that would otherwise be excluded.

Henicosafluoroundecanoic acid (C_{11}-PFCA), also known as perfluoroundecanoic acid, is a synthetically manufactured, long-chain alkane that comprises a group of chemicals known as perfluorocarboxylic acid (PFCA) (ECHA 2012a). Used in the manufacture of numerous consumer goods, C_{11}-PFCA has been identified in samples of carpeting, apparel, textiles, upholstery, floor waxes, nonstick cookware, dental floss, and many other products (Gou et al. 2009). PFCAs have been increasingly detected in household dust samples in Japan, Canada, and the United States, and long-chain PFCAs have been identified in animal tissue samples (including those of Arctic wildlife) in increasing concentrations throughout the past 30 years (EC and HC 2006; Gou et al. 2009). In August 2012, Germany issued a proposal to classify C_{11}-PFCA as vPvB under REACH. Based on WOE provisions in Annex XIII, the Member State Committee determined C_{11}-PFCA to be a vPvB in December 2012 (ECHA 2012b).

The committee evaluated all available data using expert judgment in the WOE determination to list C_{11}-PFCA as a vPvB. Information regarding C_{11}-PFCA was obtained from standardized testing, monitoring, and modeling methods; category and analog approaches (e.g., grouping and read-across); and estimation techniques (e.g., QSARs). Because of a lack of data on C_{11}-PFCA and the fact that it has not yet been registered under REACH, a read-across approach based on comparable PFCA homologue data and both abiotic and biotic degradation studies across all environmental media were considered in assessing the persistence of C_{11}-PFCA. The very great stability of fluorinated carbons (because of their high bond energies) as well as the agreement that longer-chained PFCA homologues exhibit higher boiling and melting points and lower polarizability than their counterparts led the Member State Committee to conclude that C_{11}-PFCA was at least as persistent as shorter-chained PFCAs (for which reliable data existed). Further, estimation and experimental values of C_{11}-PFCA conformed to data for shorter-chained PFCAs, where abiotic and biotic degradation pathways were largely absent at relevant environmental conditions. This led to the conclusion that C_{11}-PFCA was indeed "very persistent" (ECHA 2012b).

Committee members evaluated both field and laboratory data to assess the bioaccumulation potential of C_{11}-PFCA. BCFs for C_{11}-PFCA in whole-body systems satisfied the PBT bioaccumulation cutoff value of 2000, but not the vB cutoff value of 5000. The bioaccumulation factors (BAFs) for whole-body systems, however, all exceeded 5000, indicating that C_{11}-PFCA may be vB. Further, the BMFs of C_{11}-PFCA calculated from field studies were greater than 1, signaling the potential for biomagnification. Finally, the trophic magnification factor and terrestrial bioaccumulation assessments indicated a high potential for bioaccumulation.

Thus, even though the BCFs did not exceed the vB cutoff value, previous laboratory studies revealed hepatotoxicity in rodents exposed to PFCAs, implying that the liver is prone to bioaccumulation of PFCAs. Although this information is generally not considered in bioaccumulation assessments, the committee concluded that whole-body system studies did not reflect C_{11}-PFCA's full bioaccumulation potential, and a determination was made based on WOE and expert judgment that C_{11}-PFCA exceeded both the bioaccumulative and vB criteria under REACH (ECHA 2012b).

This case illustrates the importance of WOE in identifying chemicals of concern that may not have been identified through use of cutoff values in standard PBT tests. The German proposal to classify C_{11}-PFCA as vPvB drew attention to the chemical, indicating that Member States can play a notable role in influencing the collection of data and prioritization of chemicals under REACH.

2. Case study: Siloxane D5

Siloxane D5 (decamethylcyclopentasiloxane) is an example of a substance for which authorities in different jurisdictions have come to opposite conclusions using WOE. While Canadian authorities have concluded that Siloxane D5 does not meet the criteria for classification as a PBT, European experts have recommended that the commission treat Siloxane D5 as a vPvB. The two determinations are not necessarily at odds with one another, as Canada does not use the term *vPvB*, but the scientific reasoning of the two jurisdictions was not identical.

Siloxanes are named (e.g., D4, D5) according to their number of repeating cyclical dimethlysiloxane $(SiO(CH_3)_2)$ units. They comprise a group of industrial chemicals with widespread application in cleaning products, cosmetics, and pharmaceuticals. Employed in the manufacture of silicone polymers and copolymers, siloxanes are used in cosmetics and personal care products, including shampoos, moisturizers, and antiperspirants, to name a few. Additionally, siloxanes are used in dry cleaning, industrial defoaming applications of detergents and cleaners, and in paints, adhesives, sealants, and protective coatings. In Canada, the estimated use of Siloxane D5 in personal care products constituted 3300 tonnes in 2010 (Siloxane D5 Board of Review 2011).* In the EU, the total use of Siloxane D5 is considered confidential; a reported 17,300 and 2283 tonnes were used in personal care products and in the production of silicone polymers, respectively, in 2004 (ECHA 2012c).

Health and Environment Canada recommended Siloxane D5 for addition to the Toxic Substances List under the Canadian Environmental Protection Act in January 2009 based on the determination that it "was

---

* 1 tonne is equivalent to 1 metric ton or 1000 kg.

entering the environment in a quantity or concentration or under conditions that may have an immediate or long-term harmful effect on the environment or its biological diversity" (Siloxane D5 Board of Review 2011, p. 16). Following this determination, the industry association Silicones Environmental, Health, and Safety Council (SEHSC) of North America filed a request (July 2009) that an independent board be established to review the risks posed by uses of Siloxane D5 in Canada. SEHSC claimed that the recommendation to add Siloxane D5 to the Toxic Substances List was based on inadequate screening assessments that did not use the best available scientific data and analyses. Further, SEHSC contended that a new risk assessment should be conducted because Siloxane D5 did not meet the criteria for toxicity. The Canadian Minister of the Environment responded by assembling a board of review consisting of professional toxicologists and chemists to reassess Siloxane D5's classification as a toxic substance (Siloxane D5 Board of Review 2011).

The board reviewed the available scientific information on Siloxane D5 and ultimately concluded in 2011 that the substance does not pose a danger to the environment for both current and future uses. Although the board did not use the phrase "weight of evidence," the assessment reflects a WOE approach.

The board based its conclusion on an evaluation of the full range of studies regarding the persistence, bioaccumulation, and toxicity of Siloxane D5 as well as its uses and potential for exposure to humans and the environment. The review found that Siloxane D5 does in fact meet the criteria to be classified as persistent. However, because of its high vapor pressure and volatility, Siloxane D5 will likely partition into air, regardless of the environmental medium into which it is released. Once in the air, Siloxane D5 undergoes rapid degradation through the process of indirect photolysis (Siloxane D5 Board of Review 2011). The board concluded, therefore, that persistence was not a relevant concern without bioaccumulation.

Notably, BCF estimates for Siloxane D5 vary as to whether the 5000 cutoff value is exceeded. Evaluating all of the evidence as well as inconsistent positions of Environment Canada on the issue, the board ruled that "the values for BCF are equivocal and do not support a conclusion that the regulatory threshold for bioconcentration has been met" (Siloxane D5 Board of Review 2011, pp. 48–49). Moreover, the board found that Siloxane D5 does not biomagnify and that concentrations of Siloxane D5 in Canada's environment are at a quasi-steady state. That is, concentrations of Siloxane D5 in the Canadian environment have not grown but rather have remained relatively constant over the long-term (more than 30 years); and at current rates of use, concentration levels are not projected to rise.

Finally, a full range of toxicological assessments on both land- and aquatic-based organisms failed to provide evidence that Siloxane D5 is toxic up to its limit of solubility. On the basis of the totality of evidence, the board concluded that "it is virtually impossible for Siloxane D5 to accumulate to sufficient concentrations to produce adverse effects to organisms in air, water, soils, or sediments" (Siloxane D5 Board of Review 2011, pp. 13–14).

ECHA's PBT Expert Group and the United Kingdom's Environment Agency also evaluated Siloxane D5 to determine if it is a PBT or vPvB, and they arrived at a somewhat different conclusion (ECHA 2012c). The ECHA Expert Group reviewed a variety of standardized tests, monitoring studies, modeling methods, and estimation techniques to assess the substance's persistence, bioaccumulation, and toxicity characteristics. On persistence, the group evaluated data regarding abiotic and biotic degradation pathways, adsorption, and distribution modeling studies. Bioaccumulation assessments included estimated and measured data in aquatic species and sediment-dwelling organisms. Toxicity evaluations incorporated both aquatic plants and animals and, unlike the Canadian review board, assessments of human health.

In light of the large body of available data, the ECHA Expert Group used a WOE approach in accordance with REACH Annex XIII. The group concluded that Siloxane D5 is very persistent in sediment and, based on BCFs, vB in fish (ECHA 2012c). The group's assessment of biomagnification studies, some of which indicate that Siloxane D5 can biomagnify in certain food chains, led it to conclude that biomagnification of Siloxane D5 is inconclusive. The group concluded, however, that Siloxane D5 does not meet the toxicity criteria because the available aquatic and mammalian laboratory studies did not indicate relevant levels of toxicity. Nonetheless, the group noted, "there are some uncertainties relating to the limited available data on mammalian, avian and fish reproductive effects, and toxicity has been observed in sediment and soil organisms" (ECHA 2012c, p. 2). The Expert Group ultimately recommended in November 2012 that the substance be treated as a vPvB.

Given the limited uses of Siloxane D5 that present a risk of release to the environment, the United Kingdom Competent Authority recommended the investigation of use-specific restrictions under REACH rather than the official classification of Siloxane D5 as a SVHC. As of this writing, ECHA's registry indicates that the registrants continue to disagree with the conclusion of the EU Member States and maintain that Siloxane D5 is not a PBT or vPvB.

The Siloxane D5 case reveals how WOE can work in practice in different jurisdictions to yield conflicting results: Canadian authorities declassified Siloxane D5 as a PBT while European authorities recommended that the substance be treated as a vPvB. The Canadian Board of Review and the EU Expert Group came to opposite conclusions regarding biomagnification. Moreover, the ECHA Expert Group found much more uncertainty in the toxicity of Siloxane D5 than the Canadian review board did.

The two conclusions may not be as different as they appear on their face value. Both the ECHA Expert Group and the Canadian review board recognized that Siloxane D5 is not a PBT in that it does not meet all three criteria. Additionally, both groups recognized that the persistence of Siloxane D5 exceeds the cutoff value, and their determinations on BCFs are not altogether dissimilar, although the Canadian review found more uncertainty in BCF estimates than did the European assessment. The primary difference is in the findings regarding toxicity. However, the Canadian evaluation was informed by the limited potential for exposure in the Canadian environment, and Canada does not employ the vPvB construct. The determination that Siloxane D5 is not likely to have toxic effects therefore carries more weight in the Canadian context than in the European context. The vPvB construct exists for the very purpose of accounting for some of the uncertainty in toxicity assessments.

3.   Case study: Lindane (gamma-hexachlorocyclohexane)

Lindane, the gamma isomer of hexachlorocyclohexane (HCH), is an organochlorine insecticide. In the United States it was registered by the US Department of Agriculture in the 1940s, and many European countries produced technical HCH and lindane beginning in the 1950s. In addition to use as an insecticide for agriculture products, lindane is used to treat ectoparasites found on humans and animals (World Health Organization [WHO] 1991). The pharmaceutical application of lindane in humans is for the treatment of head lice and scabies (ATSDR 2005).

Many countries phased out lindane production between the 1970s and 1990s (United Nations Environment Programme [UNEP] 2006), although its production and use continued in some regions. Because of concerns about water quality, California banned the sale of all pharmaceutical lindane products in 2002 (Humphreys et al. 2008); four years later, US EPA canceled all agricultural registrations of lindane products. In June 2005, Mexico proposed that lindane be added to Annex A (chemicals slated for elimination) of the Stockholm Convention. In its fourth meeting, the Conference of the Parties added lindane to Annex A, effectively banning its use in agriculture, but providing an exemption for use as a pharmaceutical to treat head lice and scabies in humans (COP 2009).

Depending on environmental conditions, lindane can partition into water, air, and soil; its half-life also depends on the media and conditions. In air, estimates of the half-life range from approximately 2 to 96 days (Mackay, Shiu, and Ma 1997; Brubaker and Hites 1998). In water, half-life estimates range from 3 to 30 days and from 30 to 300 days in rivers and lakes, respectively. In soil, half-life estimates have been reported from a few days up to 3 years (WHO 1991; Mackay, Shiu, and Ma 1997). US EPA has calculated the half-life of lindane to be 110 years in the Arctic Ocean at a pH of 8, demonstrating that half-life is very dependent on environmental conditions, such as temperature and pH. Hydrolysis and photolysis have not been found to play a significant role in degrading lindane, but volatilization is important in the distribution of the chemical (WHO 1991; UNEP 2006). A joint report by WHO and the Convention Task Force on the Health Aspects of Air Pollution concluded that HCH is present all over the globe in the water and air. Evidence indicates that lindane travels through the atmosphere and, when rainfall or dry deposition occur, lindane contaminates surface soil and water in remote locations. As use of lindane has decreased during the past few decades, lower concentrations have been detected in water and air samples (Task Force 2003).

Lindane has a high lipid solubility and bioconcentrates readily, allowing it to bioaccumulate through the food chain. However, the BCFs reported for lindane are highly variable. Laboratory experiments produced BCF values from 10 to 6000, while in the field, values ranged from 10 to 2600 (WHO 1991; UNEP 2006). The BCF also varies considerably within aquatic species, and the results depend on what type of fish is used or which body parts are assayed (US EPA 2002). A review of studies on lindane concluded that bioconcentration by plants and terrestrial organisms is limited (ATSDR 2005). Conversely, UNEP (2006) states that bioaccumulation has been reported "for most taxonomic groups, from plants and algae to vertebrates" (p. 12). Lindane and other HCH isomers have also been found in human breast milk, with the beta isomer being the most prevalent (Nair et al. 1996; ATSDR 2005). Of concern is the combination of bioaccumulation with the low levels of lindane required for acute toxicity in humans and animals (UNEP 2006). There is also evidence to suggest that lindane is a potential mutagen and carcinogen (WHO 1991; ATSDR 2005).

According to the Stockholm Convention's Risk Profile for lindane, the decision to add lindane to Annex A was influenced by the overwhelming field data indicating that lindane is persistent, bioaccumulative, and toxic and has properties conducive to long-range transport. However, the field data were not necessarily consistent with laboratory data. In fact, in the risk profile for lindane, UNEP (2006, p. 20) stated:

> The evaluation of laboratory experimental data of lindane would suggest a lower potential of bioaccumulation and biomagnification than that expected for other organochlorine pesticides. In fact, lindane should be considered a border case in terms of its potential for bioaccumulation. Fortunately, there is a large amount of monitoring data on biota allowing a real estimation of the risk profile of lindane in comparison with other organochlorine pesticides. The information provided by this huge amount of real field data is conclusive: lindane concentrations

> in biota samples collected far away from use areas
> is similar to that observed for other organochlorine
> pesticides, confirming the concern for persistence,
> bioaccumulation and long-range transport.

In this case, the decision to list lindane as a POP under the Stockholm Convention relied almost exclusively on field data and biomonitoring studies, neither of which are considered standard data for PBT determinations. It is interesting that the POP Review Committee considered the field data as providing "a real estimation" of the risk profile for lindane when the field and laboratory data were in conflict. This deference to field data rather than laboratory tests was possible largely because of the extensive literature and data available for lindane, a situation that does not exist for many chemicals. This raises the following question: How extensive must field data be to override evidence from laboratory tests? Guidance on such questions would strengthen the use of WOE for PBT determinations.

***

Taken together, the three case studies demonstrate that WOE reasoning is already occurring in PBT assessment processes. They also suggest that the flexibility to consider WOE can facilitate more insightful determinations, but the danger of subjectivity and unpredictability is real. More guidance about how to apply WOE thinking in PBT processes can minimize the danger while allowing all relevant science to be evaluated.

II.   *How many PBTs are in commerce?*

One might expect the number of PBTs in commerce to be known with some degree of certainty, given that governmental regulation of chemicals has, to varying degrees, been ongoing since the 1960s in many parts of the world. In reality, however, the number of PBTs in commerce is difficult to determine and uncertainty in the number of PBTs is high. The main difficulty is the lack of measured data for most chemicals and the subsequent necessity of using estimation methods.

Reducing the uncertainty in this number is important for at least two reasons. First, the amount of time and resources that agencies must commit to identifying and regulating PBTs is largely a function of the number of PBTs in commerce. When the number of PBTs is uncertain, agencies may direct too few resources toward PBTs, resulting in risk to humans and the environment. Conversely, if too many resources are directed toward PBTs, other important initiatives may suffer. Second, lists of PBTs inform consumers about the prevalence of these substances in the marketplace. Public pressure and market forces can drive development of safer alternatives to PBTs, but only if the public is informed about the occurrence of

PBTs in consumer products. Determining the number of PBTs in commerce is therefore not merely an academic exercise but a question of practical importance in the marketplace.

The Chemical Abstracts Services (CAS) Registry includes over 70 million unique chemical substances, of which about 296,000 have been inventoried or regulated by some government around the world.* Considering industrial chemicals, pesticides, cosmetics, and pharmaceuticals, there are roughly 100,000 chemicals in commerce in developed countries, but only about 30,000 substances are produced at a quantity of greater than 1 tonne (1000 kg) per year (Muir and Howard 2006). How many of these chemicals, though, are actually of concern for human and ecological health and how many are PBTs?

By the early 2000s, fewer than 20 PBTs had been identified, primarily those identified as POPs under the Stockholm Convention. The Canadian government began categorizing the chemicals on its Domestic Substances List in 1999. That list included 23,000 substances that were manufactured or imported into Canada between January 1, 1984, and December 31, 1986. The initial screening identified 393 potential PBTs, but 145 were identified as no longer being in commerce in Canada, leaving a total of 248 potential PBTs. This example highlights the fact that lists of PBTs are often specific for a particular geographic region, and therefore, a given list might omit known or suspected PBTs because those chemicals are not relevant in that geographic (or jurisdictional) context.

Lists of PBT substances in the United States, Canada, Europe, Japan, and elsewhere vary in the number of chemicals they include (see Table 3.1). The variation among lists is explained by several factors. First, the organizations that develop the lists have differing scopes of consideration. For example, the Stockholm Convention considers only organic compounds capable of long-range transport; it lists 22 such PBTs. Second, some organizations restrict their consideration to specific geographic regions (e.g., the Canadian example above), uses, or criteria. Third, the number of chemicals assessed and the approach to developing the lists vary among organizations. Some organizations conduct direct assessment of chemicals, whereas others simply accumulate chemicals from the lists created by other organizations. California has the largest list of PBTs because it includes all substances identified as a PBT by most other authoritative organizations. California also includes heavy metals on its list of PBTs, whereas some other jurisdictions explicitly exclude metals from their consideration of PBTs. Fourth, the cutoff values are not consistent among the organizations. To determine bioaccumulation, for instance, US EPA uses a BCF of 1000 (screening purpose only), REACH uses a BCF of 2000, and the

---

* The CAS Registry is maintained by the American Chemical Society, http://www.cas.org
/content/chemical-substances.

Table 3.1 Number of PBTs identified by selected organizations

Law with PBT list	Number of PBTs	Notes
REACH (EU): substances of very high concern	20	Does not include metals Includes PBTs and vPvBs
OSPAR Convention: chemicals for priority action	42	Includes chemicals that are of an equivalent level of concern as PBTs Includes metals
Toxics Release Inventory (USA): PBTs	20	16 PBTs and 4 chemical compound categories Includes metals
Chemical Substances Control Law (Japan): Class I Substances	30	Includes metals in assessment process
CEPA 1999 (Canada): Potential PBiTs	393	Potential PBiTs based upon screening risk profiles
Stockholm Convention (UNEP): Persistent Organic Pollutants	23	Includes long-range transport criteria Does not include metals

Stockholm Convention uses a BCF of 5000 (Scheringer et al. 2012). Thus, a given chemical may be considered a PBT by some organizations but not by others.

A number of investigators have screened the inventories of existing chemicals for PBT properties. The screening methods vary, and the studies tend to list "potential" PBTs because most of the data come from estimation models rather than direct measurements. Despite differences in methodology, the studies generally find that there are likely several hundred PBTs in commerce. Below, we review several of the studies that have attempted to estimate the number of potential PBTs.

Scheringer et al. (2012) examined 93,144 industrial chemicals using the POP screening criteria from Annex D of the Stockholm Convention. This starting pool of chemicals was narrowed from an initial list of approximately 122,000 chemicals by removing 20,000 inorganic compounds, salts, and organometallic chemicals. Congeners of large groups of chemicals (e.g., polychlorinated biphenyls) were also removed. Another 10,000 organic chemicals were removed because those substances were outside the domain of the estimation methods used in the study. Of the final group of 93,144 chemicals, about half had been preregistered under REACH. Using US EPA's EPI Suite estimation tools, the authors identified 574 chemicals that potentially exceed the thresholds for persistence, bioaccumulation, toxicity, and long-range transport in Annex D of the Stockholm Convention. The authors also identified 193 halogenated chemicals that they classify as "very POP" (Scheringer et al. 2012). Using chemicals for which both measured data and estimations were available, the authors conducted an uncertainty analysis by regressing the estimated data against the measured data and determining the range of values captured within 68% (1 standard deviation from the mean) of the data points. The uncertainty analysis indicated that the lower and upper bounds for the number of potential POPs were 191 and 1201, respectively. The authors concluded that as many as 100 potential POPs could be expected for future evaluation under the Stockholm Convention. It is important to note that only aquatic bioaccumulation and aquatic toxicity were considered, so the estimate of 100 potential POPs is likely conservative. Also, the criteria for persistence and bioaccumulation in Annex D of the Stockholm Convention are less inclusive than the criteria now being used in Europe under REACH Annex XIII.

Strempel et al. (2012) examined existing and new chemicals to estimate the number of potential PBT substances on the market and to determine if the number of PBTs entering the market had declined over time. As in Scheringer et al. (2012), this study began with a much larger population of chemicals, which was then reduced by eliminating congeners, inorganic compounds, organometallic compounds, salts, and chemicals with a high molecular weight (>1000 g/mol). The authors also eliminated 1487 carboxylic and sulfonic acids for which a consistent determination of bioaccumulation potential was not possible because of discrepancies between measured log K_{ow} and the BCF estimated by EPI Suite. Finally, the authors also removed all chemicals for which the half-life was estimated solely on the basis of molecular weight (a situation known to give inaccurate results for persistence). This filtering reduced the initial pool of 127,281 chemicals to 94,483 chemicals.

Persistence, bioaccumulation, and toxicity were equally weighted in this analysis and given subscores based upon the biodegradation half-life (for persistence), BCF (for bioaccumulation), and chronic no observed effect concentration or acute median lethal concentration or half-maximal effective concentration (for toxicity). The authors searched more than ten databases to acquire as much empirical data as possible but found that only 91 chemicals had the full set of measured data (i.e., half-life values, BCF, and toxicity). Therefore, the authors used EPI Suite to estimate persistence, bioaccumulation, and toxicity for most of the chemicals.

The analysis revealed that 2930 (3.1%) of the 94,483 chemicals could be classified as potential PBTs based on the criteria presented in REACH Annex XIII. At the time of the analysis (March 2012), only 32 of the 2930 chemicals had been registered under REACH. The study also identified 1202 chemicals as potentially vPvB, nearly all of which exceeded the toxicity criterion as well. The authors conducted an uncertainty analysis that revealed lower and upper bounds for potential PBTs of 153 and 12,493, respectively. The authors concluded that several hundred PBTs were likely on the market.

In addition to the screening exercise, Strempel et al. (2012) investigated trends in PBTs entering the market and found that the number of PBTs entering into commerce may have *increased* over time. The European List of Notified Chemical Substances lists chemicals that entered into the marketplace between 1982 and 2007. Of the chemicals that entered the market during that period, 5.2% were identified as potential PBTs, whereas the percentage of potential PBTs for the entire dataset is 3.1%. This study therefore suggests that there was no downward trend in the number of PBTs being introduced into commerce between 1982 and 2007, the year that REACH entered into force.

Howard and Muir (2010) published a screening study of 22,263 chemicals gathered from the Canadian Domestic Substance List and the USEPA Inventory Update Rule database. This screening exercise addressed only persistence and bioaccumulation, and it focused primarily on commercial chemicals. As in the studies described previously, a lack of measured data necessitated the use of estimation methods, primarily EPI Suite. Unlike other studies, though, this screening exercise relied extensively on expert judgment, particularly for review of QSAR/QSPR (quantitative structure–property relationships) results. A chemical was considered potentially able to bioaccumulate if its log K_{ow} was greater than 3, or greater than 2 if its log K_{oa} was between 5 and 12. Persistence was determined largely on the basis of chemical structure and expert judgment. The authors concluded that of the nonpolymer, organic chemicals registered in Canada and the United States, more than 600 are potentially persistent and able to bioaccumulate. Of those, 101 had been measured in environmental media, either as the parent compound or as a degradation product, at the time of the study.

Rorije et al. (2011) created a scoring system for persistence and bioaccumulation and applied it to 64,721 substances gleaned from a variety of lists (some of which include pesticides). In this study, a chemical was given a P score and B score, both of which were then centered, scaled, and transformed. The persistence and bioaccumulation scores were then combined into a single "PB score." The study also used BAFs (rather than BCFs) and P_{ov} rather than the more widely used half-life values. (The authors claim that BAF and P_{ov} are better indicators of potential for bioaccumulation and persistence than are BCF and half-life measurements.) The study relies very heavily on QSARs and, therefore, the results are not applicable to inorganic compounds, organometallic substances, salts, high-molecular-weight compounds, highly reactive compounds, surfactants, and polymers. Based upon the PB-scoring method, 1986 chemicals exceed the Stockholm Convention's persistence and bioaccumulation criteria and REACH's vPvB criteria and 4541 would satisfy REACH's persistence and bioaccumulation cutoff values. The study did not assess toxicity, but toxicity is generally assumed for vPvB chemicals. If even 10% of the

4541 chemicals exceeded the REACH toxicity cutoff, this study suggests that there could be several hundred to nearly 2000 potential PBTs.

Numerous other studies have examined the relationship between different QSARs (e.g., Öberg 2006; Zachary and Greenway 2009) or have addressed potential for long-range transport (e.g., Brown and Wania 2008; Zarfl, Scheringer, and Matthies 2011). Most of these screening studies employ estimation methods, including QSARs and modeling programs such as EPI Suite, to overcome the lack of measured data for the vast number of chemicals that are assessed. The use of expert judgment to supplement and verify information from QSARs and QSPRs is common and necessary because the models are not infallible (Muir and Howard 2006). On the one hand, the use of expert judgment reduces the uncertainty and error that come with extensive reliance on estimation models. On the other hand, human judgment can introduce subjectivity, and different experts may draw conflicting conclusions from the same information. Strempel et al. (2012) determined that the largest source of uncertainty in the estimated number of PBTs was uncertainty in persistence data, suggesting that investment in measuring persistence could help resolve the number of PBTs in commerce.

Ultimately, the screening studies show that there are potentially hundreds of PBTs in commerce, many of which are not high-production-volume chemicals. Many chemicals initially identified as potential PBTs may turn out not to be PBTs after more thorough study (i.e., they are false-positives). Indeed, in some cases, fewer than 20% of the chemicals identified in screening to be potential PBTs are in fact confirmed as PBTs when studied in more detail (Davies 2013). Nonetheless, it appears that the number of PBTs currently in commerce is likely in the hundreds, with a few studies suggesting the number could approach 1000.

References

ATSDR (Agency for Toxic Substances and Disease Registry). 2005. "Health Effects." In *Toxicological Profile for Alpha-, Beta-, Gamma-, and Delta-Hexachlorocyclohexane*, 27–171. Atlanta, GA: US Department of Health and Human Services, Public Health Service.

Borgert, C.J., E.M. Mihaich, L.S. Ortego, K.S. Bentley, C.M. Holmes, S.L. Levin, and R.A. Becker. 2011. "Hypothesis-Driven Weight of Evidence Framework for Evaluating Data within the US EPA Endocrine Disruptor Screening Program." *Regulatory Toxicology and Pharmacology* 61:185–91.

Brown, T.N., and F. Wania. 2008. "Screening Chemicals for the Potential to be Persistent Organic Pollutants: A Case Study of Arctic Contaminants." *Environmental Science & Technology* 42:5202–9.

Brubaker, W.W., and R.A. Hites. 1998. "OH Reaction Kinetics of Gas-Phase α- and γ-Hexachlorocyclohexane and Hexachlorobenzene." *Environmental Science & Technology* 32:766–9.

Burkhard, L.P., J.A. Arnot, M.R. Embry, K.J. Farley, R.A. Hoke, M. Kitano, H.A. Leslie et al. 2012. "Comparing Laboratory and Field Measured Bioaccumulation Endpoints." *Integrated Environmental Assessment and Management* 8:17–31.

Burton, G.A., P.M. Chapman, and E.P. Smith. 2002. "Weight-of-Evidence Approaches for Assessing Ecosystem Impairment." *Human and Ecological Risk Assessment* 8:1657–73.

COP (Conference of the Parties to the Stockholm Convention on Persistent Organic Pollutants). 2009. Listing of Lindane. Decision SC-4/15. Fourth Meeting, Geneva, May 4–8.

Davies, E. 2013. "Searching for PBTs/vPvBs." *Chemical Watch*. Global Business Briefing, March.

EC (Environment Canada) and HC (Health Canada). 2006. "Action Plan for the Assessment and Management of Perfluorinated Carboxylic Acids and Their Precursors." *Canada Gazette Part I* 140:1542–6.

ECHA (European Chemicals Agency). 2010. *Practical Guide 2: How to Report Weight of Evidence*. Helsinki, Finland: ECHA. Available at http://echa.europa.eu /documents/10162/13655/pg_report_weight_of_evidence_en.pdf.

ECHA (European Chemicals Agency). 2012a. *Agreement of the Member State Committee on the Identification of Henicosafluoroundecanoic Acid as a Substance of Very High Concern*, December 13. Available at http://echa.europa.eu/documents/10162 /a6b5d648-3b56-4bf2-a3ce-428205151e53.

ECHA (European Chemicals Agency). 2012b. *Member State Committee Support Document for Identification of Henicosafluoroundecanoic Acid as a Substance of Very High Concern because of Its vPvB Properties*, December 13. Available at http:// echa.europa.eu/documents/10162/e359141e-e5cf-4ddf-b197-7701ea563b0f.

ECHA (European Chemicals Agency). 2012c. *Results of Evaluation of PBT/vPvB Properties of Decamethylcyclopentasiloxane*. Available at http://echa.europa .eu/documents/10162/13628/decamethyl_pbtsheet_en.pdf.

ECHA (European Chemicals Agency). 2014. *Guidance on Information Requirements and Chemical Safety Assessment, Chapter R.11: PBT/vPvB Assessment. Version 2.0*. Helsinki, Finland: ECHA. Available at http://echa.europa.eu/documents /10162/13632/information_requirements_r11_en.pdf.

Gou, Z., X. Liu, K.A. Krebs, and N.F. Roache. 2009. *Perfluorocarboxylic Acid Content in 116 Articles of Commerce*. Research Triangle Park, NC: National Risk Management Laboratory, Office of Research and Development, US EPA. EPA/600/R-09/033. Available at http://nepis.epa.gov/Adobe/PDF /P100EA62.pdf.

Howard, P.H., and D.C.G. Muir. 2010. "Identifying New Persistent and Bioaccumulative Organics among Chemicals in Commerce." *Environmental Science & Technology* 44:2277–85.

Hull, R.N., and S. Swanson. 2006. "Sequential Analysis of Lines of Evidence—An Advanced Weight-of-Evidence Approach for Ecological Risk Assessment." *Integrated Environmental Assessment and Management* 2:302–11.

Humphreys, E.H., S. Janssen, A. Heil, P. Hiatt, G. Solomon, and M.D. Miller. 2008. "Outcomes of the California Ban on Pharmaceutical Lindane: Clinical and Ecologic Impacts." *Environmental Health Perspectives* 116:297–302.

Mackay, D., W.Y. Shiu, and K.-C. Ma. 1997. *Illustrated Handbook of Physical-Chemical Properties of Environmental Fate for Organic Chemicals*. Boca Raton, FL: CRC Press.

Muir, D.C.G., and P.H. Howard. 2006. "Are There Other Persistent Organic Pollutants? A Challenge for Environmental Chemists." *Environmental Science & Technology* 40:7157–66.

Müller, T. 2014. "Screening Criteria for PBTs/vPvBs." *Chemical Watch*. Global Business Briefing, June.

Nair, A., R. Mandapati, P. Dureja, and M.K.K. Pillai. 1996. "DDT and HCH Load in Mothers and Their Infants in Delhi, India." *Bulletin of Environmental Contamination and Toxicology* 56:58–64.

Öberg, T. 2006. "Virtual Screening for Environmental Pollutants: Structure–Activity Relationships Applied to a Database of Industrial Chemicals." *Environmental Toxicology and Chemistry* 25:1178–83.

Rauert, C., A. Friesen, G. Hermann, U. Jöhncke, A. Kehrer, M. Neumann, I. Prutz et al. 2014. "Proposal for a Harmonised PBT Identification across Different Regulatory Frameworks." *Environmental Sciences Europe* 26:9.

Rhomberg, L.R., L.A. Bailey, and J.E. Goodman. 2010. "Hypothesis-Based Weight of Evidence: A Tool for Evaluating and Communicating Uncertainties and Inconsistencies in the Large Body of Evidence in Proposing a Carcinogenic Mode of Action—Naphthalene as an Example." *Critical Reviews in Toxicology* 40:671–96.

Rorije, E., E.M.J. Verbruggen, A. Hollander, T.P. Traas, and M.P.M. Janssen. 2011. "Identifying Potential POP and PBT Substances: Development of a New Persistence/Bioaccumulation-Score." Bilthoven, The Netherlands: National Institute for Public Health and the Evironment. RIVM Report 601356001/2011. Available at http://www.rivm.nl/bibliotheek/rapporten/601356001.pdf.

SCENIHR (Scientific Committee on Emerging and Newly Identified Health Risks). 2012. *Memorandum on the Use of the Scientific Literature for Human Health Risk Assessment Purposes—Weighing of Evidence and Expression of Uncertainty*. Brussels: European Commission. Available at http://ec.europa.eu/health/scientific _committees/emerging/docs/scenihr_s_001.pdf.

Scheringer, M., S. Strempel, S. Hukari, C.A. Ng, M. Blepp, and K. Hungerbuhler. 2012. "How Many Persistent Organic Pollutants Should We Expect?" *Atmospheric Pollution Research* 3:383–91.

Selin, H. 2010. *Global Governance of Hazardous Chemicals: Challenges of Multilevel Management*. Cambridge, MA: MIT Press.

Siloxane D5 Board of Review. 2011. "Report of the Board of Review for Decamethlycyclopentasiloxane (D5)." Ottawa, ON, Canada.

Solomon, K.R., M. Matthies, and M. Vighi. 2013. "Assessment of PBTs in the European Union: A Critical Assessment of the Proposed Evaluation Scheme with Reference to Plant Protection Products." *Environmental Sciences Europe* 25:1–17.

Strempel, S., M. Scheringer, C.A. Ng, and K. Hungerbühler. 2012. "Screening for PBT Chemicals among the 'Existing' and 'New' Chemicals of the EU." *Environmental Science & Technology* 46:5680–7.

Task Force (Joint WHO/Convention Task Force on the Health Aspects of Air Pollution). 2003. "Hexachlorocyclohexanes." In *Health Risks of Persistent Organic Pollutants from Long-Range Transboundary Air Pollution*. Copenhagen, Denmark: WHO Regional Office for Europe.

UNEP (United Nations Environment Programme). 2006. *Report of the Persistent Organic Pollutants Review Committee on the Work of Its Second Meeting: Addendum: Risk Profile on Lindane: Nairobi, Kenya*. UNEP/POPS/POPRC.2/17/Add.4.

US EPA. 2002. "Registration Eligibility Decision for Lindane: Case 315," September 25. Available at http://envirocancer.cornell.edu/turf/pdf/lindane_red.pdf.

US EPA. 2014. "Risk Assessment for Carcinogens." Last modified October 7. Available at http://www2.epa.gov/fera/risk-assessment-carcinogens.

Weed, D.L. 2005. "Weight of Evidence: A Review of Concept and Methods." *Risk Analysis* 25:1545–57.

WHO (World Health Organization). 1991. *Lindane (Gamma-HCH) Health and Safety Guide*. International Programme on Chemical Safety. Health and Safety Guide No. 54. Geneva: WHO. Available at http://www.inchem.org/documents/hsg/hsg/hsg054.htm.

Zachary, M., and G.M. Greenway. 2009. "Comparative PBT Screening Using (Q) SAR Tools within REACH Legislation." *SAR and QSAR in Environmental Research* 20:145–57.

Zarfl, C., M. Scheringer, and M. Matthies. 2011. "Screening Criteria for Long-Range Transport Potential of Organic Substances in Water." *Environmental Science & Technology* 45:10075–81.

chapter four

International and regional PBT policies

I. Introduction to PBT policies

Once it has been determined that a chemical is a PBT (or a potential or partial PBT), the question then becomes what decisions should be made in the public and private sectors in response to that determination. We refer to these decisions as "PBT policies."

A. Study approach to PBT policies

In this chapter, we review international and regional policies. Chapters 5 and 6 review national and subnational policies, respectively.

International and regional policies, characteristic of international environmental agreements, tend to be aspirational and/or limited in their binding content, as they reflect the collective will of many sovereign nations (Victor 1997; Barrett 2003). It is usually difficult to find an international consensus for a highly ambitious and restrictive policy. The European Union (EU) Registration, Evaluation, Authorization, and Restriction of Chemicals (REACH) Regulation is an exception as it is quite ambitious and restrictive and it is a regional initiative that relies in various ways on activities in the 28 EU member states. National policies tend to be more comprehensive than subnational policies. Along with subnational policies, we also evaluate the actions of nongovernmental organizations (NGOs) and the private sector in Chapter 6.

Although we discuss policies at each jurisdictional level (international to local), our review is not an exhaustive list of PBT policies. Among the policies we do evaluate, we do not afford equal attention to each, as some offer more insight into the principles and approaches we seek to highlight. Our review is also limited by the data that we gathered from interviews, the availability of primary texts and secondary literature in English, and resource limitations. We focus on international regimes; national efforts in Japan, China, Canada, Europe, and the United States; and subnational efforts by several US states, Ontario, NGOs, and the private sector.

The purpose of Chapters 4, 5, and 6 is to survey how the PBT concept is being utilized in various policy settings. We draw attention both to commonalities and differences. Of particular interest are differences because they may suggest a need for harmonization. We present and discuss our findings and recommendations in Chapter 7.

B.   *Policy typology and risk management principles revisited*

In Chapter 1, we introduced a policy typology to categorize PBT policies for evaluation. Policies generally take on one of three forms. First, governments and stakeholders examine PBT properties to *set priorities* for chemicals to either undergo further assessment or risk management. Second, governments may subject PBTs to higher *information requirements* than other chemicals, usually in the form of data requirements from manufacturers. Third, governments may apply *direct risk management* by discouraging, restricting, controlling, or prohibiting the use of chemicals determined to be PBTs. These broad policy types are not particular to any level of governance. Rather, we find examples of each type at every level of governance.

In addition to the policy typology, we also identified in Chapter 1 seven risk management principles. Where applicable, we explain how a particular PBT policy may reflect or fail to reflect one or more of these management principles. The risk management principles are as follows:

1. *Information quality*: The scientific information relied upon by industry and regulators should meet high standards of clarity and quality.
2. *Precaution*: When there is uncertain evidence of a potential threat to human health, safety, and the environment, lack of full scientific evidence of adverse effects should not be used as an excuse to postpone the implementation of cost-effective protective measures.
3. *Priority setting*: Decision makers should establish a systematic priority-setting process to ensure that the most undesirable risks are given appropriate priority.
4. *Exposure and risk assessment*: Before making risk management decisions about a specific technology, decision makers should be informed of a technology's potential for exposure and risk, in addition to whether or not the technology has the potential to be hazardous.
5. *Differentiation*: If the same technology is likely to have different risks in different real-world applications, the risk management and regulatory processes should be discerning enough to distinguish low-risk from high-risk applications.
6. *Rational alternatives*: When decision makers consider innovative substitutes for a potentially hazardous technology, they should also consider anticipated risks, costs, and benefits of the substitute technologies.
7. *Value of information*: Research investments should focus on projects where the value of additional information justifies the costs.

C. International and regional PBT policies

In this chapter, we review international and regional PBT policies. Both constitute governance regimes agreed to by multiple sovereign nations. Here, we simply use the terms *international* and *regional* to denote difference in geographical reach. The international regimes apply globally, while the regional regimes have more limited applicability defined by geographic/jurisdictional boundaries.

There are over 50 global and regional agreements on chemical safety as well as about 40 programs and initiatives that provide support for the agreements (Buccini 2004; Bengtsson 2010), although not all of these agreements and programs utilize the PBT concept. The international regimes we review are the Strategic Approach to International Chemicals Management (SAICM) and the Stockholm Convention on Persistent Organic Pollutants. The regional policies we consider are the Great Lakes Binational Toxics Strategy, the Convention for the Protection of the Marine Environment of the North-East Atlantic, and the EU REACH Regulation.

II. International PBT policies

A. Strategic Approach to International Chemicals Management

The 2002 World Summit on Sustainable Development (WSSD) established, as one of its goals, "to achieve, by 2020, that chemicals are used and produced in ways that lead to the minimization of significant adverse effects on human health and the environment..." (UN 2002, ¶ 23). As a precursor to the WSSD, the United Nations Environment Programme (UNEP) issued a decision in 2002 to make it a priority to improve and coordinate international chemical management activities (UNEP Governing Council 2002). To that end, governments around the world organized the first meeting of the International Conference on Chemicals Management in 2006, where the parties adopted SAICM. SAICM is an aspirational, nonbinding agreement that provides guidance to international organizations, national and subnational governments, and industry and public interest stakeholders in their development and implementation of policies for the sound management of chemicals.

The Dubai Declaration on International Chemicals Management adopts the SAICM Overarching Policy Strategy (OPS)*—a nonbinding international framework, which serves "as an umbrella mechanism to guide different management efforts as it outlines a plan of action toward

---

* SAICM, Dubai Declaration on International Chemicals Management, Art. 12. For the SAICM texts, including the Dubai Declaration, the Overarching Policy Strategy, and the Global Plan of Action, see UNEP (2006).

fulfilling the 2020 goal formulated at the WSSD" (Selin 2010, p. 5).* The OPS establishes voluntary risk reduction objectives, including the goal that by 2020:

> chemicals or chemical uses that pose an unreason-
> able or otherwise unmanageable risk to human
> health and environment based on science-based
> risk assessment and taking into account the costs
> and benefits as well as the availability of safer sub-
> stitutes and their efficacy, are no longer produced or
> used for such uses.†

The OPS further suggests that the groups of chemicals that are the highest priorities for assessment include PBTs, very persistent, very bioac-cumulative chemicals (vPvBs), and persistent organic pollutants (POPs) (as well as chemicals produced in high production volume, CMRs, and chemicals with side dispersive uses, among others).‡

The Dubai Declaration also recommends§ (but does not adopt) the SAICM Global Plan of Action, which provides guidance for SAICM imple-mentation by listing several hundred activities that particular stakehold-ers can take to achieve sustainable chemical management. In particular, the Global Plan of Action includes three actions that pertain to the gover-nance of PBTs (as well as POPs, CMR toxins, and endocrine disruptors).¶ First, SAICM promotes the substitution of hazardous chemicals with safer alternatives. Second, the approach suggests that PBTs and other hazard categories be prioritized for risk assessment above other categories of chemicals. Third, SAICM encourages parties to take an overall integrated approach to chemical management that accounts for goals and obliga-tions under international agreements and coordinates efforts with other jurisdictions.

These suggested actions, along with other actions and goals expressed within SAICM and by the WSSD, influence many of the chemical policies (including PBT policies) around the world. Elements of policies in Japan, Canada, the EU, and the United States, as well as efforts by industry and NGO stakeholders, are explicitly designed to achieve the WSSD 2020 goal (Ditz 2007; Ditz and Tuncak 2014). For example, the government of Canada has established the same 2020 date as its goal for completing efforts under its Chemicals Management Plan for assessing and managing priority chemicals.

---

* See also UNEP (2006) and Bengtsson (2010).
† SAICM, Overarching Policy Strategy, Art. 14(d)(i).
‡ SAICM, Overarching Policy Strategy, footnote 3.
§ SAICM, Dubai Declaration on International Chemicals Management, Art. 13.
¶ SAICM, Global Plan of Action, Table B, Activities 54–56.

SAICM is not a formal, binding, or self-executing risk management regime. Its goals and the guidance provided to achieve them, however, demonstrate the range of policy types. SAICM encourages priority setting, the generation and sharing of publicly available information, and management of the risks of certain chemicals or uses.

Moreover, the management principles that we identify and the recommendations that we make in this book are consistent with the PBT-related activities that SAICM promotes. In particular, SAICM emphasizes assessment of both exposure and hazard, the use of benefit–cost analysis in risk management decision making, the generation of information on priority chemicals as well as potential substitutes, and the application of the precautionary principle. Furthermore, one of the overarching objectives of SAICM is to promote harmonization and international cooperation in risk assessment methodologies as well as management approaches. In that spirit, one of the motivating ideas of this book is to highlight areas where harmonization in PBT assessment and policy might be possible.

B.   Stockholm Convention on Persistent Organic Pollutants

The Stockholm Convention on Persistent Organic Pollutants* is an international treaty that was adopted in 2001 and entered into force in 2004. The treaty is administered by UNEP and was established to protect human health and the environment from PBTs with the potential for long-range transport (Lipnick and Muir 2000; Selin 2010). Whereas SAICM constitutes a nonbinding international framework, the Stockholm Convention is a formally binding treaty that imparts risk management obligations on nations that are party to the treaty. As of 2014, 179 nations are party to the Stockholm Convention (Swedish Chemicals Agency 2013; UNEP 2014a). Afghanistan is the newest ratifying party to the convention (in 2013); the United States is a signatory but has yet to ratify the treaty.[†]

Annex D of the convention specifies the four criteria that parties must use to justify the inclusion of a chemical on one of the POP lists: persistence, bioaccumulation, potential for long-range transport, and adverse effects on human health or the environment. At a conceptual level, a POP, therefore, is a PBT that also satisfies the additional long-range transport criterion.

Any party to the convention may submit a proposal to list a chemical in one of the POP Annexes. Once proposed, the Persistent Organic Pollutants Review Committee (POPRC) applies screening criteria established

---

[*] Stockholm Convention on Persistent Organic Pollutants, May 22, 2001, 40 I.L.M. 532 (entered into force May 17, 2004) (hereinafter Stockholm Convention).

[†] Even though it has not ratified the Stockholm Convention, the U.S. has restricted the chemicals listed in the Convention.

in Annex D to determine if the substance is a POP. If a chemical meets the screening criteria, the POPRC then examines the substance through a risk profile. Annex E requires the risk profile to include data on production volume and sources, uses, releases, hazard assessments for endpoints of concern, monitoring, and environmental fate. Annex E(c) requires that bioconcentration factor (BCF) and bioaccumulation factor determinations must, as far as possible, be based on "measured values," that is, from laboratory tests rather than Quantitative Structure–Activity Relationships or models—except when monitoring data are judged to be sufficient. Through the risk profile, the POPRC must evaluate whether international action is needed to prevent the substance from causing likely harm to human health or the environment owing to its potential for long-range transport. The POPRC next develops a risk management evaluation that includes an exhaustive list of control options as well as socioeconomic data specified by Annex F, including a consideration of substitute chemicals. The Review Committee then makes a recommendation on whether the chemical should be listed in Annex A, B, or C of the convention (van Wijk et al. 2009).

The decision to list a substance falls to the Conference of the Parties. Chemicals listed in Annex A are subject to elimination, with limited specific exemptions. For example, polychlorinated biphenyls (PCBs) are subject to elimination under Annex A. However, PCBs are also subject to exemption from immediate elimination because they continue to be used in electrical transformers and other equipment. The parties to the convention have established 2025 as the sunset date for the exemption and 2028 as the goal for environmentally sound disposal management (COP 2009). Chemicals listed in Annex B are subject to restrictions on production and use, and the annex specifies acceptable purposes for production and use. Only dichlorodiphenyltrichloroethane (DDT) and perfluorooctane sulfonyl fluoride (PFOS) are listed in Annex B. DDT, for example, may be produced and used for control of vector-borne diseases. Finally, chemicals listed in Annex C are those that are formed or released unintentionally. Parties to the convention must implement the best available techniques and environmental practices to prevent the unintentional release of these chemicals. PCBs are also listed in Annex C along with dioxins and four other chemicals and chemical groups.

In total, the convention lists 23 chemicals and chemical groups, including the original "dirty dozen" chemicals and 11 additional chemicals (Swedish Chemicals Agency 2013; UNEP 2014b). The original 12 POPs are listed in Table 1.2 in Chapter 1, and Table 4.1 lists the names, common uses, and exemptions for the nine "new POPs" that were listed in 2009 as well as endosulfan, which was added to the convention in 2011, and the flame retardant hexabromocyclododecane (HBCDD), which was added in 2014 (Swedish Chemicals Agency 2013). At this writing, there are also four

Table 4.1 "New POPs" under the Stockholm Convention on persistent organic pollutants[a]

Name	Use	Exemptions[b]
Alpha hexachlorocyclohexane (alpha-HCH)	Classified as a pesticide and by-product; no current end-product uses	No exemptions
Beta hexachlorocyclohexane (beta-HCH)	Classified as a pesticide and by-product; no current end-product uses	No exemptions
Chlordecone	Classified as a pesticide; used as an agricultural insecticide, miticide, and fungicide	No exemptions
Hexabromobiphenyl	Classified as an industrial chemical, hexabromobiphenyl is an industrial chemical used as a flame retardant, mostly in the 1970s	No exemptions
Hexabromodiphenyl ether and heptabromodiphenyl ether (commercial octabromodiphenyl ether)	Classified as an industrial chemical; used as a flame retardant	Limited exemptions to account for articles containing the substance and to allow for recycling of materials containing the chemicals
Lindane	Classified as a pesticide and pharmaceutical agent; used as a broad-spectrum insecticide	Limited exemption for control of head lice and scabies
Pentachlorobenzene	Classified as an industrial chemical, pesticide, and by-product; was used in PCB products, in dyestuff carriers, as a fungicide, and as a flame retardant	No exemptions
Perfluorooctane sulfonic acid, its salts, and PFOS	Classified as an industrial chemical; used in production of polymers and in electronics, firefighting foam, photo imaging, hydraulic fluids, and textiles	Exemptions for use in manufacturing of certain products

(*Continued*)

Table 4.1 (Continued) "New POPs" under the Stockholm Convention on persistent organic pollutants[a]

Name	Use	Exemptions[b]
Technical endosulfan and its related isomers	Classified as a pesticide; used as a broad-spectrum insecticide	Exemptions for use in controlling certain crop-pest complexes
Tetrabromodiphenyl ether and pentabromodiphenyl ether (-BDE) (commercial pentabromodiphenyl ether)	Classified as an industrial chemical; used as an additive flame retardant	Specific exemptions to account for articles containing the chemical and recycling of articles containing the chemical
HBCDD	Classified as an industrial chemical; used as an additive flame retardant in expanded, extruded, and high-impact polystyrene as well as textile coatings	Specific exemptions for use in expanded and extruded polystyrene for building applications

[a] Additional details on each chemical are available through the Stockholm Convention website (http://chm.pops.int/Home/tabid/2121/Default.aspx) and the POPs Toolkit (http://www.popstoolkit.com/about.aspx).

[b] Most exemptions are effective for five years from the date of entry, and most of these exemptions entered into force on August 26, 2010 (technical endosulfan entered into force on October 27, 2012).

chemicals and chemical groups under review for inclusion in the convention, including short-chain chlorinated paraffins, chlorinated naphthalenes, hexachlorobutadiene (HCBD), and pentachlorophenol (UNEP 2014c). In 2014, Norway submitted a proposal to add bis(pentabromophenyl) ether (deca-BDE) to the convention; the Conference of the Parties will make a decision on deca-BDE and other chemicals in 2017 (European Chemicals Agency [ECHA] 2014l, p. 45).

The convention uses the PBT concept for all three policy types. First, the convention is limited in scope in that it applies only to PBTs with the potential for long-range transport. In effect, this utilizes the PBT concept as a priority-setting mechanism to identify the chemicals that, because of their properties, pose risk on an international basis. Second, the convention facilitates information gathering on chemicals, including their PBT properties as well as their uses, exposure pathways, and potential substitutes. The convention itself does not directly compel industry to provide or generate data; however, it does place an obligation on nations that are party to the treaty to provide, share, and evaluate information. To fulfill these obligations, Article 10(3) directs parties to gather information from industry. Finally, the convention applies risk management by listing chemicals in its annexes for elimination, restrictions, and best practices management for unintended production and releases.

Overall, the design of the POP regulatory process appears to be consistent with the management principles. The convention places a premium on clear, transparent, and accurate information. Article 9 of the convention specifies obligations relating to international information exchanges. Annex E demonstrates a preference for data from laboratory tests and observational studies, when available, rather than estimation techniques. The convention also has a solid foundation in the precautionary principle that scientific uncertainty should not constitute a barrier to cost-effective risk management (Godduhn and Duffy 2003; Maguire and Ellis 2005). The precautionary principle is cited in Article 1 of the convention as part of its objective, and Article 8(9) instructs the Conference of the Parties to make listing decisions "in a precautionary manner." Additionally, the convention uses the persistence, bioaccumulation, and toxicity criteria along with potential for long-range transport as a priority-setting tool: A substance must satisfy those criteria to be considered for risk management. However, a POP determination does not trigger an automatic regulatory outcome, as the convention requires additional risk assessment and socioeconomic analysis, including consideration of production volume, releases, the availability and risks of chemical substitutes, and alternative risk management options (RMO) for specific uses.* Thus, the convention is consistent with the exposure and risk assessment principles and the rational alternatives principle. The convention also adheres to the differentiation principle by allowing for phased elimination for particular applications under Annex A and for continued specific uses under Annex B where the benefits of use outweigh the risks, such as DDT for vector control.

Ultimately, the Stockholm Convention constitutes the primary embodiment of PBT policy applied at a global level. The convention demonstrates what we find to be the primary use of the PBT concept in the policies that we review: that PBT properties are first assessed to identify priority chemicals of concern, and risk assessment and management are then applied as appropriate.

III. Regional PBT policies

A. Great Lakes Binational Toxics Strategy and the Great Lakes Water Quality Agreement

The governments of Canada and the United States established the Great Lakes Water Quality Agreement (GLWQA) of 1972, administered by the

---

* Stockholm Convention, Art. 9(1).

International Joint Commission (IJC), in response to eutrophication in the Great Lakes caused primarily by phosphorus runoff (AGL et al. 2007). In 1978, the governments expanded the scope of the agreement to address other concerns, including pollution from industrial chemicals (AGL et al. 2007). The 1978 Agreement, amended in 1987, established the goal of "virtual elimination" of discharges of persistent toxic pollutants (initially most of the dirty dozen) into the Great Lakes Basin (Williams 2006).* The 1987 Agreement also designated chemicals with all three PBT characteristics as "critical pollutants" and encouraged the development of management plans for these chemicals.† The governments of Canada and the United States developed the Great Lakes Binational Toxics Strategy of 1997 out of this process. Recently, the 1987 Agreement and the Binational Toxics Strategy have been superseded by the GLWQA of 2012. For the purposes of understanding the evolution of PBT policy, it is important to outline the Binational Toxics Strategy.

The Binational Strategy was a collaborative, voluntary effort between Environment Canada (EC) and the US Environmental Protection Agency (US EPA) (1997). As in the GLWQA of 1972, the goal of the Binational Strategy was the virtual elimination of discharges of persistent toxic pollutants into the Great Lakes Basin. Note that the goal of virtual elimination of discharges focuses stakeholders' efforts on preventative measures. On the other hand, under Canada's Toxic Substances Management Policy of 1995, virtual elimination refers to the elimination of anthropogenic PBTs in discharges as well as the remediation of such substances already present in the environment (EC 1995; Williams 2006). The Binational Strategy therefore embraced a somewhat limited variant of the virtual elimination concept. As explained below, the Canadian Environmental Protection Act of 1999 adopts a similar definition of virtual elimination, where elimination refers to a quantity of release that is "below the level of quantification."‡

The Binational Strategy formally addressed persistent toxic substances through two lists. Level I substances included those that had been previously listed by US EPA and EC individually and by the IJC as critical pollutants. All Level I substances were persistent and toxic, but not necessarily bioaccumulative, although most satisfied all three PBT criteria (EC and US EPA 1997, Annex I; Williams 2006). They included mercury, PCBs, dioxins and furans, hexachlorobenzene, benzo(a)pyrene,

---

* Great Lakes Water Quality Agreement, US–Can., Nov. 22, 1978, 30 U.S.T. 1383; Protocol Amending the Agreement on Great Lakes Water Quality, US–Can., Nov. 18, 1987, T.I.A.S. No. 11,551 (hereinafter 1987 GLWQA) (amending 1978 Great Lakes Water Quality Agreement). The agreement defines virtual elimination as "zero discharge" of persistent toxic pollutants. 1987 GLWQA, Art. II(a); Annex 12(2)(a)(ii).
† 1987 GLWQA, Art. IV(1)(f)(2), Annex 2(1)(b).
‡ Canadian Environmental Protection Act of 1999, § 65(1).

octachlorostyrene, alkyl-lead, chlordane, aldrin/dieldrin, DDT, mirex, and toxaphene.

Level II substances were those for which either EC or US EPA had evidence that indicated that the substances satisfied all three PBT criteria, but the evidence has not yet been sufficiently considered by both nations such that they could reach an agreement on joint challenge goals. There were 13 Level II chemicals and one group of chemicals (polycyclic aromatic hydrocarbons [PAHs]). The agreement provided for periodic reconsideration of Level II substances for inclusion on the Level I list.

To accomplish its objectives, the Binational Strategy facilitated information gathering about point and nonpoint sources of pollution within the Great Lakes Basin. The focus was on identifying how each Level I substance is used and released from each source, the efficacy and progress of existing programs that manage or control the substances, and the cost-effective options to achieve further reductions.

Finally, for each Level I substance, the Binational Strategy set challenge goals to guide stakeholders toward achievement of the overall goal of virtual elimination of discharges. As the strategy was not a formal treaty, the challenge goals were not binding. Rather, the strategy encouraged information exchange, coordination of efforts, and voluntary management by stakeholders (e.g., voluntary emission reductions from pollution sources or the enactment of risk management by Great Lakes states and provinces or national governments). As of 2009, 13 of the original 17 challenge goals for Level I substances had been met (EC and US EPA 2009).

In 2012, the Canadian Minister of the Environment and the Administrator of the US EPA signed a protocol to amend the GLWQA, which replaces the 1987 GLWQA and the Binational Toxics Strategy.* The GLWQA of 2012 entered into force in February 2013. Like the prior agreements, it is not legally binding. Rather, it provides a framework for cooperation between federal, provincial, state, and local governments in the Great Lakes Basin. The agreement maintains the overall virtual elimination goal and establishes new general and specific objectives. General objectives include the provision of "safe, high-quality drinking water" from the Great Lakes, "consumption of fish and wildlife unrestricted by concerns due to harmful pollutants," and that the Great Lakes should "be free from pollutants in quantities or concentrations that could be harmful to human health, wildlife, or aquatic organisms, through direct exposure or indirect exposure through the food chain."† The protocol also calls for the development of ecosystem objectives for each Great Lake and

---

* Protocol Amending the Agreement on Great Lakes Water Quality, U.S.–Can., Sept. 7, 2012 (hereinafter 2012 GLWQA), available at http://binational.net/wp-content/uploads /2014/05/1094_Canada-USA-GLWQA-_e.pdf.
† 2012 GLWQA, Art. 3(1)(a).

substance objectives for those substances that present a potential for harm to the Great Lakes ecosystem.*

The GLWQA of 2012 encourages parties to develop, in cooperation with stakeholders, pollution abatement, control, and prevention programs for hazardous substances (as well as for agricultural runoff, invasive species, and other risks to environmental wellbeing).† Annex 3 to the protocol addresses pollution from industrial chemicals by facilitating the identification of chemicals of mutual concern to the parties and obliging the parties to prepare binational strategies to address harm from those chemicals, which may include the development of water quality standards, the reduction of releases, promotion of safer substitutes, monitoring, and exchange of information.‡

While the agreement does not specify criteria for identifying chemicals of mutual concern, it seems reasonable to believe that persistence, bioaccumulation, and toxicity will play a role. The first set of chemicals of mutual concern includes many of the chemicals on the Binational Strategy lists, including mercury, PCBs, nonylphenol, chlorinated paraffins, polybrominated diphenyl ethers (PBDEs), HCBDs, and perfluorinated compounds. Bisphenol A is also listed as a chemical of mutual concern, although it raises concern because of endocrine disrupting effects, not PBT-related concerns (Galatone 2014). These chemicals and chemical groups were selected as the pilot chemicals for the GLWQA of 2012 because they are national priorities in Canada and the United States, they present known or suspected risks to the environment and human health, and environmental data are readily available.

Overall, the Binational Toxics Strategy used PBT determinations to set priorities for information gathering and to guide risk management by industry and authorities at different levels of government. Like the Canadian Environmental Protection Act of 1999, a primary goal pertains to the virtual elimination of discharges. The GLWQA of 2012 maintains the virtual elimination goal as well as a goal of zero discharges for chemicals of mutual concern. However, the GLWQA of 2012 is not necessarily a PBT policy because the agreement does not specify that persistence, bioaccumulation, and toxicity should be used as properties for identifying chemicals of mutual concern.

Furthermore, any actions that the parties take to achieve the virtual elimination objective are voluntary, as the agreement does not impart legal obligations on the parties that would be enforceable under international law. As a framework to coordinate regulatory efforts and to guide information sharing, the United States and Canada (and states

---

* 2012 GLWQA, Art. 3(1)(b).
† 2012 GLWQA, Art. 4.
‡ 2012 GLWQA, Annex 3.

and provinces) will rely on their own policies to fulfill commitments. For example, the government of Canada will primarily utilize the Canadian Environmental Protection Act of 1999 and the Chemicals Management Plan as the vehicles to fulfill Canada's commitments under the GLWQA of 2012 (Galatone 2014).

As of yet, it is difficult to know if the GLWQA of 2012 adheres to the risk management principles, as it is in its early stages of implementation. The former Great Lakes agreements helped guide priority setting for government and stakeholders, and the GLWQA of 2012 seems to be fulfilling the same role. In Chapter 5, we explain and evaluate the national policies of Canada and the United States, and in Chapter 6, we comment on Ontario's PBT policy.

B.   Oslo–Paris Convention for the Protection of the Marine Environment of the North-East Atlantic

The Oslo–Paris (OSPAR) Convention, which entered into force in 1998, is an agreement among 15 countries in Northwestern Europe and the EU to protect the marine environment of the North-East Atlantic.* To that end, the Convention includes annexes that address prevention and elimination of pollution from land-based sources, dumping or incineration, and off-shore sources, as well as an annex on environmental quality assessment. To fulfill the mandates of the convention and its annexes, the OSPAR Commission developed the North-East Atlantic Environment Strategy, which encompasses five thematic strategies: biodiversity, eutrophication, radioactive substances, offshore oil and gas industry, and hazardous substances that present risks to ecosystems in the North-East Atlantic.

The PBT concept is incorporated into the annexes and the Hazardous Substances Strategy. OSPAR defines "hazardous substances" as those that are persistent, bioaccumulative, and toxic, as well as substances of "equivalent concern" (OSPAR Commission 2010a; Moermond et al. 2012). Annex I on pollution from land-based sources and Annex III on pollution from off-shore sources identify PBTs (and substances of equivalent concern) as targets for release reduction and phase-out.[†] The objective of the Hazardous Substances Strategy is to reduce discharges of hazardous substances "with the aim to achieve concentrations in the marine environment near background values for naturally occurring substances and close to zero for manmade synthetic substances" (OSPAR Commission 2010a, Part II, Hazardous Substances, ¶ 1.1).

---

* Convention for the Protection of the Marine Environment of the North-East Atlantic Sept. 22, 1992, 2354 U.N.T.S. 67 (hereinafter OSPAR Convention).
† OSPAR Convention, Annex I, Art. 3(a); Annex III, Art. 10(b).

To identify priority substances for assessment and management, in 2001, the OSPAR Hazardous Substances Committee commissioned an informal group of experts to utilize a decision-making process tool called the "Dynamic Selection and Prioritisation Mechanism for Hazardous Substances" (DYNAMEC) (OSPAR Commission 2006). The mechanism involved three stages of assessment to focus regulatory decision making under OSPAR. First, substances were screened (based on existing experimental and modeling data) for possible PBT characteristics as well as "safety net" criteria for endocrine disrupting chemicals, metals, inorganic compounds, and other substances that are found in food webs in the marine environment. This screening process generated a list of substances that are eligible and noneligible for regulation under OSPAR. The list was validated by an informal group of experts, and a decision was made by OSPAR's Hazardous Substances Committee to place chemicals on a List of Substances of Possible Concern. Second, chemicals on the list were ranked based on an exposure score (considering production volume, use patterns, and monitoring data) and an effect score (based on indirect and direct effects on marine ecosystems). Third, the OSPAR Commission made a decision based on the ranking to place chemicals on a List of Chemicals for Priority Action.

Using relatively permissive cutoff values to define persistence and bioaccumulation (see Table 2.1), the DYNAMEC process initially led the OSPAR Commission to list some chemicals that were subsequently deselected after the submission of additional information from industry and an expert review of the cutoff values used for defining persistence, bioaccumulation, and toxicity. Ultimately, eight substances were removed from the List of Chemicals for Priority Action and 77 were removed from the List of Substances of Possible Concern. At present, there are 315 chemicals or chemical groups on the List of Substances of Possible Concern and 42 on the List of Chemicals for Priority Action (OSPAR Commission 2015a).

For hazardous substances on the List of Chemicals for Priority Action, the OSPAR Commission may issue recommendations or apply legally binding risk management on parties. Under OSPAR, there have been about 60 recommendations and legally binding risk management decisions to regulate point sources of discharges as well as diffuse sources, including use of hazardous substances in products. For example, the OSPAR Commission has promoted best available techniques for emission reductions and best environmental practices to achieve specified limit values for heavy metals, organohalogens, and PAHs. It has issued regulations pertaining to specific industries, including the manufacturing of textiles, chlorine, organic chemicals, pulp and paper, and others. It has also applied various degrees of risk management to specific uses of substances, including tributyltin, PAHs, nonylphenols, short-chain chlorinated paraffins, and PFOS (OSPAR Commission 2010b, 2015b).

Although 42 chemicals remain on its priority list, the convention now relies primarily on REACH and a host of EU laws (e.g., the Water Framework Directive)* to regulate hazardous substances (OSPAR Commission 2015a).† The OSPAR Convention is nonetheless important in the history of PBT regulation because it is the source of some of the initial thinking about highly inclusive cutoff values for persistence, bioaccumulation, and toxicity.

The convention adheres to many of the risk management principles we identify. OSPAR is another example of the embodiment of the precautionary principle, including its application of inclusive cutoff values to define persistence and bioaccumulation.‡ Additionally, the focus on discharges of hazardous substances is a good example of emphasizing the evaluation of exposure in risk assessments. The DYNAMEC ranking process also incorporates information about exposure and use into its listing processes. The convention is also consistent with the priority-setting principle as it identifies persistence, bioaccumulation, and toxicity as prioritization criteria for assessment of land-based and offshore pollution sources.§ The Hazardous Substances Strategy also suggests that parties should be guided by the substitution principle (emphasizing the adoption of lower risk substitutes for priority chemicals) and the principle that risk assessment should be used for setting priorities in addition to being a basis for management decisions (OSPAR Commission 2010a).

C. EU: Registration, evaluation, authorization, and restriction of chemicals

Several EU legislative instruments address PBTs. Here, we evaluate the REACH Regulation.¶ With technical administration by the ECHA in Helsinki, Finland, and policy decisions by the European Commission in Brussels, Belgium, REACH is a complex compilation of several bodies of regulation that govern the cradle-to-grave manufacture, importation, sale, and use of industrial chemicals. As EU law, REACH constitutes a regional policy; however, it shares many elements in common with national programs.

---

* Council Directive 2000/60/EC, Framework for Community Action in the Field of Water Policy, 2000 O.J. (L 327) 1 (EU).
† The OSPAR Convention still enforces monitoring of the marine ecosystem for PBTs.
‡ OSPAR Convention, Art. 2(2)(a).
§ OSPAR Convention, Appendix 2, paragraph 1.
¶ Parliament and Council Regulation 1907/2006, 2006 O.J. (L 396) 1 (EC) (hereinafter REACH). Other programs that pertain to PBTs include the following: Commission Regulation 850/2004, Persistent Organic Pollutants, 2004 O.J. (L 158) 7 (EC) (hereinafter POPs Regulation), and Council Regulation 528/2012, Biocidal Products, 2012 O.J. (L 167) 1 (EU).

As its name suggests, the REACH Regulation is composed of four disparate, but connected, programs: registration, evaluation, authorization, and restriction. Under *registration,* virtually all new and existing chemicals manufactured and sold in Europe above 1 tonne per year must be registered with ECHA. Under *dossier evaluation,* ECHA ensures, through auditing, that industry-authored registration dossiers are complete and valid. Under a complementary *substance evaluation* process, Member States coordinated by ECHA assess the risks associated with a small number of chemicals of concern. REACH targets "substances of very high concern" (SVHCs) such as CMRs and PBTs for potential risk management. After authorities identify potential SVHCs, they determine which regulatory vehicle would be most suitable for a particular chemical use, including the *authorization* and *restriction* processes under REACH or processes under other legislation (e.g., directives related to chemical exposures in the workplace). Under REACH, SVHCs are placed on a Candidate List to denote that they are candidates for authorization. They may then be listed on the Annex XIV Authorization List. Chemicals on the Authorization List are subject to mandatory phase-out and substitution, unless industry meets strict evidentiary requirements through a use-specific authorization process. The EU can also initiate restrictions on specific uses of chemicals to address "unacceptable risk."*

The PBT concept is incorporated into REACH in many ways. Using weight of evidence evaluation under Annex XIII, the REACH legislation calls for special attention to PBTs and a closely related group of substances that are vPvB, for which authorities are unable to definitively evaluate toxicity through standard methods. REACH Article 57 and Annex XIII specify the criteria for identifying SVHCs, and Article 59 specifies the process for doing so. An SVHC is a PBT, vPvB, CMR, or substance of equivalent concern. In identifying and regulating SVHCs, REACH utilizes the PBT and vPvB designations for each of the identified policy types in our typology: priority setting as well as a variety of legally binding information and management obligations on government and industry.

In particular, REACH employs the PBT concept in the following ways (ECHA 2015a):

1. Registration
    a. If a firm manufactures or imports a substance at 10 tonnes or more per year, its registration dossier must include a Chemical Safety Report (CSR), including a PBT assessment. If a registrant

---

* For a description of the REACH regulatory processes, see Bergkamp (2013) and Abelkop et al. (2012). ECHA guidance documents are available online at http://echa.europa.eu /support/guidance.

finds that a chemical is a PBT or vPvB, then the registrant must include additional information on risk management in the dossier, and the registrant and downstream users must make good faith efforts to minimize exposure and release of the chemical throughout its lifecycle.

b. In evaluating proposals for animal testing, ECHA must give priority to potential SVHCs, including potential PBT and vPvB substances.

2. Evaluation
   a. Under dossier evaluation, ECHA uses PBT and vPvB properties, among other criteria, to prioritize registration dossiers for auditing.
   b. Under dossier evaluation, ECHA evaluates registrants' PBT assessments to ensure compliance with regulatory and scientific standards.
   c. Under substance evaluation, Member States may conduct PBT assessments as part of the substance evaluation process.

3. Authorization
   a. Candidate List: Inclusion of an SVHC on the Candidate List triggers supply-chain informational requirements. In addition, if a chemical is identified as a PBT or vPvB on the Candidate List, registrants of that chemical must also recognize it as such in their registration dossiers, prompting the registration-related risk management requirements.
   b. Authorization List: Inclusion of an SVHC on the Authorization List triggers a phase-out period leading to an eventual prohibition on use, manufacture, and importation of the chemical in the EU. Industry may apply for limited, use-specific exemptions to the ban. PBTs and vPvBs are considered "nonthreshold" substances to denote that, legally, there is no safe threshold at which risks can be adequately controlled. Therefore, applications for authorized uses of PBT and vPvB substances must be based on socioeconomic justifications for their continued use.

4. Restriction
   a. The commission may place use-specific restrictions on chemicals, including PBT and vPvB substances.

5. SVHC Identification and RMO
   a. Under REACH Article 59, Member States and ECHA, at the request of the commission, may propose chemicals for identification as SVHCs, including those that satisfy the PBT or vPvB criteria under REACH Annex XIII.
   b. SVHCs, including PBTs, vPvBs, and substances of equivalent concern, may be subjected to RMO analysis to determine the most suitable option (if any) for risk management.

In the following subsections, we address each of these applications of the PBT concept in REACH in greater detail.

1. Registration

Registration is based on the principle of "no data, no market"*—the notion that nearly all chemicals on the market warrant basic safety data. The general registration provision requires that "any manufacturer or importer of a substance … in quantities of 1 tonne or more per year shall submit a registration to [ECHA]."[†] Downstream users—often small or large companies that make use of a chemical in consumer products or services—may also provide use and safety information on their own or assist in the preparation of registration dossiers through a lead registrant.[‡]

The registration dossier under REACH must contain a minimum set of data, or the substance may not be put on the market in the EU. The registration must take the form of a technical dossier, which includes information on the identity of the manufacturer, importer, or producer; the identity, including chemical and physical properties, of the substance; the manufacture and uses of the substance; environmental fate and pathways; (eco)toxicological information; guidance on safe use; and research summaries.[§]

The registration process is tiered into three phases based on firm-level production and importation volume as well as hazard characteristics. The tiers in the registration process influence the data requirements that are applicable. Chemicals produced or imported in higher volumes and chemicals that exhibit certain hazardous properties (e.g., CMR properties and aquatic toxicity) not only have earlier registration deadlines but also have more demanding data requirements.[¶]

As of December 2014, registrants have submitted 49,781 dossiers for 12,890 substances (ECHA 2014a). Of these, 49 substances and groups have been identified in registration dossiers as PBT or vPvB substances, and 66 substances require additional information to carry out a PBT assessment.

The registration process applies to PBTs (as well as vPvBs and substances of equivalent concern) in two ways. First, the registration requirements for PBT, vPvB, and equivalent concern substances are subject to additional information requirements. Once the 10-tonne production threshold is reached for a registrant, a CSR for the substance must be added to the registrant's registration dossier.[**] The CSR is the documentation of a chemical safety assessment, including information on hazards

---

* REACH, Art. 5.
† REACH, Art. 6(1).
‡ REACH, Art. 37.
§ REACH, Art. 10(1)(a).
¶ REACH, Art. 12(1) and 23.
** REACH, Art. 10(1)(b) and 14(1).

to human health and the environment, physiochemical hazards, and an assessment of whether the substance qualifies as a PBT or vPvB.* If the safety assessment reveals that the substance is hazardous or qualifies as a PBT or vPvB, then additional information is required, including exposure scenarios and risk characterization. Also, companies throughout the supply chain must implement operational conditions and risk management measures to "adequately control" or minimize the risks, including potential release and exposure.† Here, the PBT designation triggers additional risk assessment and/or management by the registrants and downstream users.

Second, REACH requires ECHA to evaluate every proposal for additional testing on vertebrate animals, to prevent unnecessary testing. Registration dossiers should include minimum data sets of risk-relevant information. If registrants do not have certain information, then they might consider tests, including animal testing, to generate the needed data. REACH seeks to limit the amount of animal testing done to meet registration requirements. Therefore, registrants must seek ECHA's approval before conducting experiments using vertebrate animals. In evaluating testing proposals, ECHA must give priority to proposals in registrations of substances that "have or may have PBT, vPvB, [CMR] properties, or substances classified as dangerous ... [and are manufactured or imported at quantities] above 100 tonnes per year with uses resulting in widespread and diffuse exposure."‡

Under registration, the PBT and vPvB designations operate under a precautionary regulatory framework that emphasizes group classifications, management based on intrinsic properties and use-specific exposure potential, preventative risk management, and reversed burden of proof (Karlsson 2010). Chemicals that fall into the PBT or vPvB hazard categories are automatically treated as suspect, subject to additional information and preventative management requirements. The "no data, no market" registration principle also reversed the burdens of proof of safety and information, shifting them from government to industry (Abelkop and Graham 2015).

Registration also advances the other principles of information quality, priority setting, exposure and risk assessment, differentiation, rational alternatives, and value of information. To be sure, there are quality deficiencies with some of the data in registration dossiers, including data for PBT assessments (e.g., Rudén and Hansson 2010; Gilbert 2011; Rovida, Longo, and Rabbit 2011; Scheringer 2013; Ball et al. 2014; Stieger et al. 2014). However, there are standards in place to encourage a minimum threshold

---

* REACH, Art. 14(3).
† REACH, Art. 14 and 31.
‡ REACH, Art. 40(1).

of data quality, to facilitate the generation of higher-quality data when necessary, and for ECHA to check the quality of data in dossiers (dossier evaluation). Further, registrants include notes on the quality of data cited in their dossiers. When additional data from animal tests are needed, suspected PBTs (and other hazard categories) are prioritized above other chemicals in evaluating the needs for the data and proposed methods. Preventative management of PBT and vPvB substances in registration is based on exposure and risk in its focus on minimizing exposures throughout a chemical's lifecycle. The emphasis on exposure here also allows companies to differentiate between different uses in the type of supply-chain management that they apply. Higher information requirements on PBTs and vPvBs also provide an incentive for companies to seek out safer alternatives that do entail use of PBTs or vPvBs.

One point of potential confusion for registrants is how REACH treats PBT and vPvB substances (as well as certain categories of CMRs) as non-threshold substances since, under REACH, the risks from the use of these substances cannot legally be considered to be "adequately controlled" for the purposes of authorization.* In other words, the legislation treats these substances as if there is no safe nonzero level of emission or exposure. However, under registration, CSRs included in registration dossiers must specify measures to adequately control risks,† and downstream users also have obligations under registration to ensure that risks are adequately controlled.‡

One way to interpret these provisions is that REACH allows for registrants to affirmatively certify that risks associated with PBTs and vPvBs are adequately controlled under registration, but once a PBT or vPvB substance has been included on the Authorization List, a case for "adequate control" of risks can no longer be made. Confusion on this point is likely, as other regulatory bodies (e.g., the US EPA in the United States) do not assume that PBTs have no safe level of release or exposure.

Finally, the registration process reflects the value of information principle since the amount of information required in the dossier is adjusted based on what will be necessary for safety evaluation. Nonetheless, we have learned that some view registration as an exercise that creates a great deal of paperwork without much material benefit in environmental or public health protection. Registration is intended to fulfill protection goals by incentivizing the generation of more information on chemicals to fill data gaps, by compelling companies throughout the supply chain to revisit and bolster risk management measures to minimize chemical

* REACH, Art. 60(2)–(3).
† REACH, Art. 14(6).
‡ REACH, Art. 37(5)–(6).

exposures, and by compelling businesses throughout supply chains to communicate with one another with regard to risk management.

Critics also tend to question the application of detailed risk assessment requirements on chemicals, or certain uses of chemicals, that are believed, based on scientific judgment, to be safe. However, there is value in certifying that chemicals, or specific uses, believed to be safe actually are safe. The resulting confidence is expected to boost the public's level of confidence in both government and industry. The ongoing challenge with registration and dossier evaluation (described below) is to ensure that information in the dossiers meets high standards of quality and is available to be used to inform decision making by businesses, consumers, and governments throughout the world.

2. *Evaluation*

REACH contains two distinct evaluation processes: dossier and substance evaluation (Herbatschek, Bergkamp, and Mihova 2013). Under dossier evaluation, ECHA evaluates a specific registration dossier. Dossier evaluation is a compliance check that is meant to verify that the industry-generated registration dossiers fulfill REACH's data requirements (ECHA 2013a). ECHA's team of experts determines whether the dossier includes the required and appropriate data. The experts then analyze the quality of the data by evaluating the reliability and validity of the studies cited in the dossier. The team also examines exposure scenarios, which are required for PBTs, vPvBs, CMRs, and all "classified" (dangerous)* substances manufactured or imported at volumes greater than 10 tonnes per year (i.e., those classified under the EU's version of the Globally Harmonized System for a hazardous property). Last, ECHA evaluates the risk management measures described in the dossier and may consider whether the actions are likely to be sufficient to achieve adequate control of risks.† The team may request more data to support the effectiveness of risk management or suggest that additional actions be considered.

Under dossier evaluation, ECHA uses the PBT determination as a priority-setting tool. REACH requires ECHA to conduct an evaluation on at least 5% of the registration dossiers it receives within each registration tonnage category (defined as firm production volume of 1000, 100, and 1 tonnes per year). Of that 5% baseline, ECHA chooses 25% of the dossiers to check at random and 75% based on "concern for safe use," that is, if the

---

* Substances classified as dangerous under Council Directive 67/548, Classification, Packaging and Labelling of Dangerous Substances, 1967 O.J. (L 196) 1 (EEC).

† REACH, Annex I § 6.4, indicates that risk is adequately controlled if the estimated exposure levels will not exceed the derived no-effect level or the predicted no-effect concentration for the substance and the likelihood and severity of an event occurring because of a physiochemical property of the substance (e.g., flammability, explosivity) are negligible.

substance is PBT, vPvB, or CMR; if the substance has a wide dispersive use; or if the dossier raises technical concerns such as excessive confidentiality claims.*

Substance evaluation is an altogether different process, carried out by Member States in collaboration with ECHA and the European Commission (ECHA 2014b). It involves evaluation of a specific substance rather than a specific dossier. It covers all uses of the substance, not simply those addressed by a registrant in a specific registration dossier. Whereas dossier evaluation is a form of compliance check and registration requires industry-generated safety assessments, substance evaluation allows authorities in Member State governments to perform more detailed risk assessments for specific chemicals of concern. The outcomes of a substance evaluation can trigger regulatory obligations under any of the other provisions of REACH as well as other EU legislation. For substance evaluation by Member States, prioritization can be based on PBT properties as well as exposure information such as uses, estimated releases, and production quantity (ECHA 2011a).†

The REACH evaluation processes fulfill the information quality principle in that they are both designed to assess the quality of information included in the registration dossiers (dossier evaluation) and on particular substances (substance evaluation) but to not waste resources by reviewing every registration dossier or every substance. The evaluation processes are also designed to allow for the generation of more information when authorities believe it is necessary or appropriate, and thus, the iterative nature of information submission is consistent with the value of information principle. Finally, each process utilizes the PBT concept to set priorities on which dossiers and substances should be assessed first. Thus, they exploit the priority-setting principle as well.

3.   Authorization

REACH uses the PBT concept in its authorization process as well. Authorization aims to minimize risks from SVHCs, which include PBTs, vPvBs, CMRs, and substances of equivalent concern.‡ Authorization is a complicated, stepwise process involving two lists. First, a substance is classified as an SVHC (we discuss the SVHC identification process later in this chapter). ECHA, at the request of the Commission, or a Member State government may request that a substance be placed on the Candidate List by submitting a dossier in accordance with Annex XV of REACH to identify

---

* ECHA also employs an information technology tool to assist in selecting which dossiers to evaluate, but there is human judgment in the selection process as well. REACH, Art. 41(5).

† REACH, Art. 44(1).

‡ REACH, Art. 57.

the substance as an SVHC (see ECHA 2014c).* Then, the SVHC is placed on a Candidate List, denoting that the chemical is a candidate for phase-out under authorization. The ECHA's Member State Committee, a committee of experts composed of representatives from the Member States, evaluates each substance that has been proposed for inclusion on the Candidate List (see ECHA 2015b). A unanimous decision of the committee places the substance on the Candidate List while a split vote turns the listing decision over to the commission.† Once the substance is on the Candidate List, the commission, through comitology, may add the substance to the Annex XIV Authorization List (Herbatschek, Bergkamp, and Mihova 2013; ECHA 2014d).‡ This process is depicted in Figure 4.1. Placement of a substance on the Authorization List is a severe action that triggers a phase-out of use, production, and importation of a chemical unless the commission authorizes specific uses that industry has applied for on a time-limited basis.

Placement of the chemical on the Candidate List triggers supply-chain informational requirements.§ Information on a chemical makes its way through the supply chain via a document called a Safety Data Sheet (SDS), which must include information on hazard identification, first-aid measures, accidental release measures, handling and storage, personal protection, transport, ecological and toxicological effects, and other factors.¶ An SDS must accompany any PBT and vPvB, any chemical designated as dangerous in accordance with the EU's classification and labeling directives, and any other SVHC. An SDS must be provided upon request to any person who receives a preparation containing a PBT or vPvB that is present in the preparation at 0.1% or greater by weight basis (ECHA 2015c).** On the basis of this information, purchasers of preparations that contain SVHCs (including PBTs or vPvBs) are able to urge their suppliers to reduce or eliminate the worrisome components, which is an illustration of the principle of rational alternatives. In addition, if a chemical is

---

*   REACH, Art. 59.
†   To date, all Member State Committee decisions on candidate listing have been unanimous, with contentious negotiation occurring before voting (see Herbatschek et al. 2013, p. 157).
‡   Comitology is the process by which the commission adopts implementing acts to apply uniformly throughout the EU without each individual Member State government having to adopt implementing legislation. Under comitology, the commission drafts an implementing act for submission to a committee of representatives of the Member States referred to as the REACH Comitology Committee (distinct from the ECHA Member State Committee). The comitology committee then decides whether an implementing act should be adopted through a majority vote. See Parliament and Council Regulation 182/2011, Laying Down the Rules and General Principles Concerning Mechanisms for Control by Member States of the Commission's Exercise of Implementing Powers, 2011 O.J. (L 55) 13.
§   REACH, Art. 31–33.
¶   REACH, Art. 31(6). The information requirements of safety data sheets are governed under the provisions of REACH, Art. 31 and 32.
** REACH, Art. 31(3).

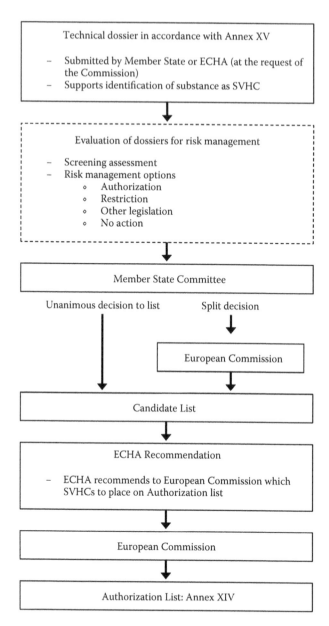

Figure 4.1 Process for inclusion of substances on the REACH Authorization List. (Versions of this table also appear in Abelkop et al. [2012] and Abelkop and Graham [2015].)

added to the Candidate List as a PBT or vPvB, registrants of that chemical must edit their registration dossiers to reflect that classification. Finally, the inclusion of a chemical on the Candidate List may also trigger some degree of market deselection of the chemical because of the stigma associated with being on a formal regulatory list and also because of the greater likelihood that the chemical may be placed on the Authorization List (e.g., Heitman and Reihlen 2007; CSES 2012; Grunwald and Hennig 2014; RTG 2015).

ECHA uses a qualitative and quantitative scoring system to determine the order in which chemicals on the Candidate List will be assessed for inclusion on the Authorization List (ECHA 2014j). The prioritization criteria are listed in Article 58(3) as those chemicals with PBT/vPvB properties, wide dispersive uses, or high volumes of production or importation. ECHA's scoring system assigns qualitative scores of "very low" to "very high" priority as well as numerical scores of 0 to 15 based on each of those criteria.

Overall, the most severe outcome of a PBT or vPvB designation under REACH is that such designation will lead to the placement of a chemical on the Annex XIV Authorization List or that it will be phased out under the restriction mechanism (which is discussed below). The purpose of the authorization process is to protect public health and the environment by compelling the substitution of SVHCs with suitable, safer alternatives (ECHA 2011b).[*] In this way, REACH embodies the principle of precaution and advances the ideals of green and sustainable chemistry (DeHihns, Hey, and Zygmont 2009; Tuncak 2013; Westervelt 2013).

Generally, registrants can obtain authorization to continue a specific use of a chemical for one of two reasons: by demonstrating that the risks of the use will be adequately controlled or by demonstrating through socioeconomic analysis that the benefits outweigh the risks for the particular use, provided that there are no existing suitable alternatives. If the chemical has an established dose–response relationship with a safety threshold, the applicant may secure authorization by demonstrating that the health and environmental risks from a specific use are "adequately controlled" relative to the threshold, as documented in the CSR.[†] However, as we note above, this adequate control provision may not be used to grant the authorization of PBTs, vPvBs, certain CMRs, and certain substances of equivalent concern. In effect, REACH, without any definitive scientific rationale, treats PBTs and vPvBs as nonthreshold substances (i.e., no safe nonzero level of emission or exposure), which is quite different from the practice in North America. Therefore, the applicant must use the alternative authorization justification by demonstrating that the socioeconomic

[*] REACH, Art. 55.
[†] REACH, Art. 60(2).

benefits of continued use outweigh the risks and that there are no suitable alternatives.* An applicant must present a substitution plan if suitable alternatives are available or a research and development plan to discover a suitable alternative if one is not readily available.† As of this writing, we know of no case where a socioeconomic rationale has been approved under REACH for continued use of a PBT or vPvB.

As of December 2014, there are 161 chemicals on the Candidate List, of which 20 are PBT, vPvB, or both (ECHA 2014e). There are currently 31 chemicals on the Authorization List. Of those, HBCDD is the only listed PBT, and musk xylene is the only listed vPvB (ECHA 2015d). Chemicals on the Authorization List remain on the Candidate List. Of the other 18 PBTs and/or vPvBs on the Candidate List, two are listed solely as PBT; two are listed as PBT and toxic to reproduction; four are listed as vPvB; four are listed as PBT and vPvB; two are listed as PBT, vPvB, and carcinogenic; and four are listed as PBT, vPvB, carcinogenic, and mutagenic. Table 4.2 lists chemicals on the Candidate List that are identified as PBT and/or vPvB, and Table 4.3 lists chemicals on the Authorization List that are identified as PBT and/or vPvB.

Several PBT-related actions were taken under REACH in late 2014. The commission added nine substances to the Authorization List, although none of these were added due to PBT-related concerns. ECHA also released draft recommendations to add 22 Candidate List substances onto the Authorization List. Among these 22 candidates, anthracene oil and coal tar pitch raise PBT and vPvB concerns because of their PAH constituents (ECHA 2014f, 2014g). The most recent PBT/vPvB additions to the Candidate List (December 2014) are two derivatives of benzotriazole (UV-320 and UV-328), which meet both the PBT and vPvB criteria.

In assessing the authorization process through the lens of our risk management principles, we show that it has beneficial aspects to it but that it is ultimately a blunt risk management instrument. To its credit, the authorization process does facilitate information gathering and review of information quality beyond that of the registration process alone. The SVHC identification, candidate listing, and authorization listing processes all consider the weight of evidence, with many opportunities for stakeholder engagement and input. The application process under authorization also facilitates the generation of use-specific data, which gives industry an opportunity to enhance public confidence in the safety of the most beneficial and significant uses of SVHCs.

The authorization process is a precautionary tool by its nature, especially in its treatment of PBTs, vPvBs, and certain other substances as nonthreshold substances. Authorization may seem to be too blunt as an

---

* REACH, Art. 60(4).
† REACH, Art. 60(4)(a)–(d).

Table 4.2 SVHCs identified as PBTs and/or vPvBs on the REACH Candidate List

Name	Classification	Prioritized for authorization
Ammonium pentadecafluorooctanoate (APFO)	Toxic for reproduction, PBT	
Pentadecafluorooctanoic acid (PFOA)	Toxic for reproduction, PBT	
Bis(pentabromophenyl) ether (decabromodiphenyl ether; DecaBDE)	PBT, vPvB	
Heptacosafluorotetradecanoic acid	vPvB	
Pentacosafluorotridecanoic acid	vPvB	
Tricosafluorododecanoic acid	vPvB	
Henicosafluoroundecanoic acid	vPvB	
Anthracene oil, anthracene paste, anthracene fraction: CAS number 91995-15-2	Carcinogenic, mutagenic, PBT, vPvB	
Anthracene oil, anthracene-low: CAS number 90640-82-7	Carcinogenic, mutagenic, PBT, vPvB	
Anthracene oil, anthracene paste: CAS number 90640-81-6	Carcinogenic, mutagenic, PBT, vPvB	
Pitch, coal tar, high temperature	Carcinogenic, PBT, vPvB	Prioritized for authorization
Anthracene oil, anthracene paste, distillation lights: CAS number 91995-17-4	Carcinogenic, Mutagenic, PBT, vPvB	
Anthracene oil: CAS number 90640-80-5	Carcinogenic, PBT, vPvB	Prioritized for authorization
Alkanes, C10–13, chloro (short-chain chlorinated paraffins)	PBT, vPvB	
Bis(tributyltin)oxide (TBTO)	PBT	
Anthracene	PBT	
HBCDD, alpha-HBCDD, beta-HBCDD, gamma-HBCDD	PBT	Authorization list
5-tert-butyl-2,4,6-trinitro-*m*-xylene (Musk xylene)	vPvB	Authorization list
2-(2H-benzotriazol-2-yl)-4,6-ditertpentylphenol (UV-328)	PBT, vPvB	
2-benzotriazol-2-yl-4,6-di-tert-butylphenol (UV-320)	PBT, vPvB	

Table 4.3 SVHCs identified as PBTs and/or vPvBs on the REACH
Authorization List

Name	Classification	Authorized uses
HBCDD, alpha-HBCDD, beta-HBCDD, gamma-HBCDD	PBT	None
5-tert-butyl-2,4,6-trinitro-*m*-xylene (Musk xylene)	vPvB	None

approach to risk management. Since PBTs are considered nonthreshold substances without scientific demonstration, the authorization process is not consistent with the exposure and risk assessment principle. To be sure, the combination of persistence and bioaccumulation makes exposure assessment difficult. Our interview data revealed conflicting opinions on the extent to which exposure assessments of PBTs could be considered valid and reliable. European authorities view persistence and bioaccumulation as exposure-related characteristics that may make accurate exposure estimates impossible, while others we interviewed tend to see the PBT designation as a hazard category. However one views the PBT category, management decision making that does not meaningfully account for how a chemical is used violates the exposure and risk assessment principle. The authorization process does respect the differentiation principle by allowing industry to make a case for continuation of some uses under stringent criteria.

In our view, placing the burden of proof on industry to defend specific uses of PBTs and vPvBs is a positive element of REACH. Authorization, as noted above, provides industry with an opportunity to continue certain uses of SVHCs by showing that the benefits of continued use justify the resulting risks. The authorization process also fulfills the rational alternatives principle to the extent that it incentivizes the consideration of safer substitutes or alternative industrial processes that altogether forego chemicals in certain applications. Authorization may therefore be a stimulus to green and sustainable chemistry.

It is difficult to state any finding on authorization with confidence because the authorization procedures under REACH have not yet been fully implemented. In December 2013, Rolls-Royce became the first company to receive an opinion from ECHA that the commission should approve authorized use of a substance on the Authorization List by making the case that risks are adequately controlled in a specific aerospace application: The seven-year authorization is for the use of DEHP—an abbreviation for bis(2-ethylhexyl) phthalate, a reproductive toxin—in the manufacture of aero engine fan blades (ECHA 2014h). In 2013, ECHA received a total of eight authorization requests covering two phthalates in 17 different uses. In 2014, ECHA received 19 authorization applications covering 38 different uses, including an application submitted by 13 coapplicants for two uses of the PBT flame retardant HBCDD in building applications (ECHA 2014i).

Before the DEHP authorization decision, the common perception among industry stakeholders was that the authorization process will be strict, onerous, and unpredictable with regard to outcome (Herbatschek, Bergkamp, and Mihova 2013). Such perceptions are likely to evolve as practical experience with the authorization process is accumulated. As of late 2014, there is no precedent for an authorization based on socioeconomic analysis; the HBCDD authorization applications will be the first, if they are approved. In the following paragraphs, we discuss the restriction process.

4. Restriction

REACH authorizes the commission to issue restrictions on the manufacture, placement on the market, or use of selected chemicals to address unacceptable risks to human health and the environment that must be addressed at the community-wide level (European Commission 2014).* Authorization applies only to placement of a chemical on the market and use. Restrictions may apply to manufacturing, uses, and the presence of a chemical in articles, including imported articles. Restrictions may apply to a single chemical or a group of chemicals (e.g., phenylmercury compounds). They may take the form of complete bans, concentration limits, prohibitions of particular uses, or limitations to particular uses. Finally, authorization applies only to SVHCs, whereas a restriction can be applied to any chemical regardless of whether it has been identified as an SVHC. Although restrictions may entail a wide variety of measures, they have generally been applied on a use-by-use basis. Restrictions are listed in REACH Annex XVII.†

When issuing a restriction, the analytic burden of proof rests with the commission. Restriction proposals may be prepared by Member States or by ECHA either at the request of the commission or for a substance on the Authorization List after the date by which use and placement of the substance on the market are prohibited. To propose a restriction, the authority must prepare an Annex XV dossier, which must include documentation of an unacceptable risk to human health or the environment, justification of action at the EU community level, assessment of the feasibility of alternative substances or technologies, assessment of the risks of alternatives, socioeconomic analysis of the proposed restriction, and RMO analysis that shows that the proposed restriction is the most appropriate measure

---

* REACH, Art. 68(1).
† As of October 2014, Annex XVII lists restrictions on 64 chemicals and chemical groups (ECHA 2015e).

considering the effectiveness, practicality, and enforceability of plausible alternative management options.

The restriction authority under REACH has essentially been carried forward from prior legislation.* Some view restrictions as the "safety net" under REACH to address risks that are not adequately addressed through registration, evaluation, and authorization (ECHA 2007). Unacceptable risk as a safety standard is not defined, providing authorities with wide discretion (Herbatschek, Bergkamp, and Mihova 2013). For PBTs, vPvBs, and substances of equivalent concern (and other non-threshold substances), unacceptable risk may be linked to the adequate control standard: Risk may be unacceptable if there are releases that are not considered adequately controlled. Under Article 69(1), a restriction proposal must be requested if authorities "consider" that a chemical presents a "risk to human health or the environment that is not adequately controlled and needs to be addressed…"† Authorities have prepared fewer restriction proposals than Candidate List proposals, presumably because they prefer to put the safety burden on industry rather than making the detailed required substantive showings needed to support a restriction proposal (Herbatschek, Bergkamp, and Mihova 2013).

As a risk management approach, the restriction mechanism provides a good illustration of how a regulatory program can adhere to the principle of differentiation of uses because of its application on a use-by-use basis. Additionally, restriction proposals must consider substitute chemicals/processes and weigh the risks of those alternatives against the risks of the target substance, thereby satisfying the rational alternatives principle. Restriction authority may adhere to the principle of exposure and risk assessment, as it is customary for the EU to support restrictions with exposure and risk assessments of specific uses. However, the undefined unacceptable risk standard may provide authorities with too much discretion if it allows them to base management decisions more on hazard while disregarding exposure considerations. By itself, the restriction authority does not adhere to the priority-setting principle because there is no prioritization scheme to guide the development of restrictions. As noted below, the SVHC identification and RMO processes may guide the application of restrictions in the coming years. However, restrictions may legally apply to more chemicals than those that have been designated SVHCs, and there is no prioritization framework to guide authorities in considering chemicals or uses for restriction.

---

* Council Directive 76/796, Restrictions on the Marketing and Use of Certain Dangerous Substances, 1976 O.J. (L 262) 201 (EEC).

† Herbatschek et al. (2013) write, "This suggests that the term 'unacceptable risk' may encompass all inadequately controlled risks, but these terms are not entirely synonymous; otherwise REACH would only have used the term 'adequately controlled' instead of 'unacceptable risk'" (p. 147).

Restrictions have not been applied to many PBT or vPvB substances. However, we discuss two cases later in this chapter—a restriction on phenylmercury compounds and a proposed restriction on deca-BDE—to show how restrictions can apply to PBTs. Before we address these cases, the following section discusses issues regarding SVHC identification and the choice of risk management approaches under REACH.

5. *SVHC identification and RMO*

Before the authorization and other risk management processes, authorities must determine which substances are SVHCs in the first place. In this process, the PBT and vPvB designations serve as priority-setting criteria. Among the nearly 100,000 chemicals in commerce in the European marketplace, only about 1500 chemicals have been identified as potential SVHCs (Herbatschek, Bergkamp, and Mihova 2013). Here, we describe the SVHC identification process that currently exists under REACH as it pertains to PBTs.

After the identification of a chemical as an SVHC, European authorities must also determine which type of risk management tool to apply, including authorization or restriction under REACH or management under different legislation (e.g., laws on pesticides or worker safety). This decision-making process about which risk management approach to take is referred to as RMO analysis.

In February 2013, the commission released the *Roadmap on Substances of Very High Concern*, (the *SVHC Roadmap*) outlining plans to identify, assess, and make risk management decisions on SVHCs through 2020 (European Commission 2013). In December 2013, ECHA released its *SVHC Roadmap to 2020 Implementation Plan*, which provides additional details on how the goals in the commission's roadmap will be fulfilled (ECHA 2013b). The *SVHC Roadmap* and ECHA's *Implementation Plan* clarify the SVHC identification, prioritization, and RMO processes under REACH.

In the following subsections, we discuss how the PBT concept is used in SVHC identification and prioritization, potential issues that may arise under the "substances of equivalent concern" category, and how the RMO fits into the REACH risk management framework.

a. SVHC identification and prioritization For PBTs, implementation of the *SVHC Roadmap* is influenced by the activities of the PBT Expert Group, which assists in identifying and recommending PBTs for classification as SVHCs. The group was established in 2012 and consists of experts from Member States, the government of Switzerland, the commission, and stakeholder organizations (e.g., an industry group called the European Chemical Industry Council and an environmental NGO called European Environmental Bureau). According to ECHA's website, "The PBT Expert Group provides informal, non-binding scientific advice on questions related to the identification of PBT and vPvB properties of chemicals" (ECHA 2015f). Annex XIII of REACH specifies the criteria for finding that a chemical is persistent, bioaccumulative, and toxic. It also requires authorities to take a weight of the evidence approach in determining whether or not a substance is a PBT. (See Chapter 3 for additional information on the weight of evidence approach.) As of December 2014, the PBT Expert Group had finalized PBT assessments for 29 chemicals—finding five to satisfy the PBT/vPvB criteria—and has 99 assessments ongoing in coordination with Member States, the commission, and ECHA (ECHA 2015g).

Based on existing classifications of substances under various EU laws (e.g., the Classification, Labeling, and Packaging Regulation),* early estimates indicated that there might be as many as 1500 substances eligible for classification as an SVHC (Herbatschek, Bergkamp, and Mihova 2013). In 2013, the commission estimated that, at most, 440 substances will have to be assessed for SVHC classification by 2020, with a most likely rate of 20 to 30 substances per year (European Commission 2013). Each SVHC may undergo a rudimentary or screening-level assessment before a management decision on how to proceed.

To determine which potential SVHCs should be studied and listed first, REACH specifies prioritization criteria for assessing chemicals on the Candidate List for authorization in Article 58(3) as PBT and vPvB characteristics, wide dispersive uses, significant market-level production and importation volume, and ECHA's capacity to deal with the authorization applications (ECHA 2014j). Those are the criteria that are expected to focus the deliberations of the PBT Working Group.

In its *SVHC Roadmap*, the Commission specified criteria for identifying "relevant SVHCs" for prioritization to undergo RMO analysis. Relevant SVHCs are those that meet the SVHC criteria listed in Article 57 (including PBTs, vPvBs, CMRs, and substances of equivalent concern), those that are registered for the nonintermediate uses, those for which the prima facie case of unacceptable risk (triggering restriction) cannot be currently made, those that are not exempt from authorization, and those that are not subject to regulation under other EU legislation (European Commission 2013). Interestingly, this list of criteria goes beyond the Article 58 criteria for determining priorities for authorization. Before the release of the *Implementation Plan*, no comprehensive, formal procedure had been specified for setting priorities among potential SVHCs, generating some stakeholder uncertainty around the prioritization process before RMO analysis (Herbatschek, Bergkamp, and Mihova 2013).

---

* Parliament and Council Regulation 1272/2008, Classification, Labelling and Packaging of Substances, 2008 O.J. (L 353) 1 (EC).

ECHA's *Implementation Plan* (2013b) outlines the process by which chemicals in commerce will be selected for RMO analysis. The registration database will constitute the primary source of information, and chemicals registered for nonintermediate uses will be prioritized for RMO analysis. First, authorities will apply an automated program to search the registration database for chemicals that potentially satisfy the criteria for Article 57 hazard and fate SVHC properties (CMRs, PBTs, vPvBs, and substances of equivalent concern). Second, authorities will apply an automated screening program to the potential SVHCs that are registered for nonintermediate uses to screen for selection criteria, including high volume, highest potential for fulfilling the criteria for Article 57 SVHC properties, structural similarity to chemicals on the Candidate List, and additional informational needs. The outcome of the screening will yield a pool of "substances of potential concern" (ECHA 2013b, p. 12). Authorities will manually screen a number of chemicals to review the efficacy of the automated screening program.

Third, screened chemicals will be sorted into various hazard groups (e.g., potential PBTs, CMRs). Expert groups within ECHA will assess the chemicals to determine if they satisfy the criteria to be considered SVHC. The PBT Expert Group is responsible for determining whether potential PBTs meet the Annex XIII criteria for classification as a PBT, vPvB, or substance of equivalent concern to a PBT/vPvB. Separate from the screening process, chemicals may also be referred to the PBT Expert Group by Member States.

Fourth, if the PBT Expert Group determines that a chemical meets the Annex XIII criteria, the chemical will be added to the pool of chemicals subject to RMO analysis. If the group determines that there is not enough information or that existing information is of too poor of quality to make a determination on the Annex XIII criteria, then the chemical may be subjected to a compliance check or substance evaluation in an effort to gain information sufficient to make a determination. The *Implementation Plan* notes that chemicals requiring additional information will be subject to further prioritization. As additional information is added to the registration database, screening will undergo regular reiterations. In addition to potential PBTs, chemicals that present other potential hazard considerations (e.g., potential CMRs) will also be subject to similar processes to determine if they are relevant SVHCs to undergo RMO analysis.

b.   Substances of equivalent concern   REACH's substance of equivalent concern category constitutes another source of uncertainty in the SVHC identification process. The commission may use the category in a variety of ways, including some ways that relate to potential PBTs. The original thinking under REACH was to focus the category primarily on endocrine-disrupting chemicals and potential PBTs. The category may also be used to include partial PBTs, substances that meet only one or two of the PBT criteria but which do not meet the thresholds set for formal designation as a PBT or vPvB.

The equivalent concern category lacks a precise definition in the legislation.* If authorities in one or more Member States believe that the chemical could be troublesome (despite not meeting the strict numeric criteria), they might seek to persuade ECHA and the commission that the chemical is a substance of equivalent concern. If there is conflict, the Member State Committee under REACH is likely to play an important role. As we discussed earlier, this equivalent concern categorization may be seen as a viable alternative to stretching the weight of evidence to support a PBT or vPvB determination.

Without opining on the legal issues, there are some drawbacks to this approach as well as benefits. Since a PBT determination entails three findings (persistence, bioaccumulation, and toxicity), it may be questionable to claim that a chemical that satisfies only two or one of the properties could reasonably be of equivalent concern. The entire point of the PBT determination is to give more priority to substances that possess all three characteristics. Moreover, if a chemical poses unacceptable risks but is not a PBT or vPvB, it can be regulated under REACH's restriction process, which may be use-specific and informed by risk assessment and socioeconomic analysis.† Allowing partial PBTs (PTs or BTs or PBs) into the SVHC listing process will dilute the priority-setting focus of the SVHC designation and make the REACH authorization process even more unpredictable for chemical manufacturers and users, who need to make fixed investments in supply chains, production facilities, and product formulations.

The methylmercury restriction discussed below somewhat illustrates this point. Authorities used the equivalent concern category to classify methylmercury (and hence phenylmercury) as a PBT-like substance. The restriction proposal found that methylmercury satisfied the bioaccumulation and toxicity requirements, but it was indeterminate based on existing evidence as to whether or not it met the persistence criteria. This is not necessarily a case where the equivalent concern category was used to classify a bioaccumulative and toxic substance as an SVHC. Rather, it seems like the equivalent concern concept was used to justify regulation where a weight of evidence determination could not be made on persistence because of a lack of data. Perhaps, where the weight of evidence is indeterminate as to whether a substance meets one of the three PBT criteria, the equivalent concern category can be used to justify a chemical's classification as a PBT-like substance rather than as a full PBT (although they are treated as regulatory equivalents).

---

* REACH, Art. 57(f).
† REACH, Art. 68(1).

Liberal use of the equivalent concern category in the REACH process of making SVHC determinations may risk the loss of some credibility both domestically and internationally, since it could appear that European regulators are prepared to bend the system on a specific chemical to achieve a predetermined result that is preferred by particular Member States. If the commission should consider addressing partial PBTs through SVHC listings, it may be wise to consult broadly with stakeholders and on an international basis to avoid unintended impacts on regulatory predictability and international trade.

c. Risk management options In addition to uncertainty surrounding the SVHC identification and authorization application processes for PBTs, there is broader uncertainty regarding the interaction of the various risk management approaches under REACH (see Bergkamp and Herbatschek 2014). Originally, the REACH Regulation was interpreted such that all substances identified as SVHCs would eventually be placed on the Candidate List, and all substances on the Candidate List would eventually be placed on the Authorization List. Hence, all SVHCs would be placed on the Authorization List. For PBTs and vPvBs, ECHA's website indicates, "PBT/vPvB substances included in the Candidate List will be subject to prioritisation and will eventually be included into Annex XIV to the REACH Regulation (the Authorisation List)" (ECHA 2015a). However, questions have arisen over the legal necessity and desirability of placing all SVHCs on the Candidate List and all Candidate List substances on the Authorization List. Over the past few years, the Commission and ECHA have developed a policy approach to address this question called the RMO analysis, which is described in the Commission's *SVHC Roadmap* and ECHA's *Implementation Plan*.

The *SVHC Roadmap* contains a proposal from the commission to modify REACH implementation by adding a new step before placing a substance on the Candidate List: an RMO assessment (European Commission 2013). A formal RMO process has not yet been established; however, the RMO proposal calls for consideration of alternatives to placing substances on the Candidate and Authorization Lists. Those alternatives may include risk management under other EU legislation or other regulatory processes within REACH (e.g., restriction). Some PBTs, such as polychlorinated terphenyls and short-chain chlorinated paraffins, are already subject to restrictions under REACH and other regulations.* The commission plans

---

* Short-chain chlorinated paraffins were previously restricted under REACH but were removed in 2013 from REACH Annex XVII and prohibited as a persistent organic pollutant the POPs Regulation (850/2004/EC) after being added to the Aarhus Protocol on Persistent Organic Pollutants (briefly mentioned in Chapter 1).

to subject all PBTs and vPvBs to RMO analysis (European Commission 2013).

Instead of making priority-setting decisions among chemicals placed on the Candidate List, the commission will prioritize relevant SVHCs for RMO analysis. As noted above, the identified criteria used to determine whether an SVHC is relevant include whether the substance meets the Article 57 SVHC criteria for Candidate Listing (including PBT properties), whether the substance is registered for nonintermediate use, whether risk is not adequately controlled (e.g., by demonstrating environmental releases in the case of PBTs) such that it triggers the restriction process, whether the specific uses are exempted from REACH's authorization provisions, and whether the known uses are already regulated by other EU legislation that supplies a "pressure for substitution" (European Commission 2013, p. 10). Although the Roadmap does not explicitly mention the additional Article 58(3) criteria for prioritizing Candidate List chemicals for inclusion on the Authorization List (PBT properties, wide dispersive uses, and high production volume), ECHA's *Implementation Plan* suggests that these criteria will also play a role in prioritizing SVHCs for RMO analysis.

The RMO process illustrates how authorization, restriction, and other EU legislation interact with one another. RMO analysis will presume that authorization is preferable to restriction, and authorities will have to overcome that presumption to conclude that an alternative risk management option is desirable (Bergkamp and Herbatschek 2014). The criteria for determining relevant SVHCs suggest that if a chemical is already subject to some other form of risk management, then it will be deprioritized for RMO analysis. If, for example, the commission decides to regulate a PBT through REACH's process of restriction, then it may avoid making the determination that the substance is an SVHC for purposes of authorization under REACH. In cases where insufficient data are available to choose the best risk management option, the commission may instead pursue a formal evaluation of the registration dossiers or a formal substance evaluation by a Member State as provided for in REACH.

From the perspective of the management principles, the RMO step is a promising innovation. By considering the extent of releases of PBTs and vPvBs before placing a chemical on the Candidate List, the RMO approach reflects a "soft" application of the principle of exposure and risk assessment. By allowing use-specific restrictions instead of the general stigmatization associated with SVHC listing, the RMO approach may protect low-release, high-benefit applications in accordance with the differentiation principle. Also, the option of deferring decisions with insufficient data to ECHA or the Member States is consistent with the information quality principle, although it may be in tension with the precautionary principle.

There is some disagreement among stakeholders about whether it is permissible for the commission to apply the RMO approach before chemicals have been placed on the Candidate List or whether RMO analysis should occur between the Candidate Listing and the Authorization Listing (EEB and ClientEarth 2012). In addition, the commission is only publishing RMO conclusions rather than the analysis itself. Greater transparency in decision-making processes is needed (Bergkamp and Herbatschek 2014). To enhance the legitimacy of the RMO approach, we suggest that the commission propose chemical-specific decisions (e.g., whether to pursue restriction or authorization) under the RMO approach for public comment, including stakeholder consultation, before reaching a final decision. While adding this additional process will delay the commission's implementation progress to some extent, the overall improvement in the credibility (and possibly the quality) of the process justifies a limited delay.

6.   Case studies

a.   Management of HBCDD through the authorization process   Given the uncertainty surrounding the authorization process, stakeholders have raised concerns as to whether the process is workable and fit-for-purpose (see ECHA 2013c). If authorizations prove to be effectively impossible to obtain in the case of PBTs and vPvBs, then the REACH Regulation, as it applies to PBTs and vPvBs, may be considered a violation of both the principles of exposure and risk assessment and differentiation of low-risk from high-risk uses. Authorization applications have been submitted for one only PBT/vPvB on the Authorization List: HBCDD. The responses from ECHA and the commission to the HBCDD authorization applications will be an important test case.

HBCDD is a brominated flame retardant. It may be identified as a mixture of all of its diastereoisomers or as a combination of its three primary diastereoisomers, which are designated alpha, beta, and gamma (US EPA 2010, 2014). It comes in the form of a white powder or pellets and is used as an additive in the manufacture of expanded polystyrene foam (EPS) and extruded polystyrene foam (XPS), which are used in the construction sector for insulation in buildings. It is also used as a flame retardant in electronics (high impact polystyrene), and it may also be micronized for use in textiles (e.g., upholstery for furniture and automobile interiors) (ECHA 2013d; US EPA 2014).

Member State and EU-level regulations establish fire safety requirements for various products, including those used in building construction (HBCDD Consortium 2013b). The manufacture of EPS and XPS, in particular, must include some flame retardant chemical. Overall, the construction sector uses HBCDD in the manufacture of roughly 0.7 million tonnes per year of EPS and 0.5 million tonnes per year of XPS for floor, wall, and roof insulation (ECHA 2013d).

To meet the demand for flame retardant additives, it is estimated that about 10–14,000 tonnes/year of HBCDD are manufactured or imported into the EU, about half manufactured in the EU and half imported (ECHA 2009a; IOM Consulting 2009; VECAP 2013).* Of this amount, about 5300 tonnes of HBCDD were used in the formulation of EPS, and 5900 tonnes were used in the formulation of XPS (ECHA 2009a, 2013d). The amount of HBCDD that is imported in articles (e.g., packaging materials, textiles, electronics) is uncertain but likely considerable (ECHA 2009a).

HBCDD is registered under REACH by eight joint registrants for 10,000 to 100,000 tonnes per year.[†] In the EU, HBCDD is manufactured at only one site in the Netherlands. In 2006, it was used by about 80 formulators of EPS and XPS, 600 EPS converters, 24 producers of textile coatings, 21 producers of XPS articles, thousands of professional users of articles containing EPS and XPS, and an unknown number of producers and recyclers of electronic articles containing high-impact polystyrene (ECHA 2009a). In its end uses, HBCDD is in products and buildings throughout Europe.

Because HBCDD does not chemically bind to polymers, it has the potential for release at all stages throughout its lifecycle from production to disposal (US EPA 2010, 2014). It was estimated in 2007 that HBCDD is released in the EU at a rate of 3 tonnes per year, with roughly 50% of releases into wastewater, 21% into air, and 29% into surface water (ECHA 2009a). However, releases of HBCDD have likely declined since 2007 (VECAP 2013). The primary sources of releases are industrial users and formulators of polystyrene and textile coatings, although HBCDD is also released during manufacturing and product service life. Monitoring data show that the presence of HBCDD in biota throughout Europe, including in the Arctic region, is pervasive (ECHA 2009a). Humans may be exposed in the workplace, through dust indoors, and through food, especially near industrial sites (Swedish Chemicals Agency 2007).

HBCDD has hazardous properties. In the 1990s, HBCDD was identified as a chemical of potential concern under the OSPAR Convention. The registration dossier on HBCDD indicates that it is a suspected reproductive toxin, that babies may be exposed through breast milk, and that it can be very toxic to aquatic life with long-term effects. In addition, the registration dossier identifies HBCDD as a PBT and vPvB substance.[‡]

---

[*] Authorization applicants use the same data source as ECHA and estimate 14,000 tonnes per year in 2011. Commenters on the authorization application critique the methodology used to generate that estimate and suggest that the appropriate range is about 10–12,000 tonnes per year (see Chemtura 2014, Appendix 2).

[†] Registration dossiers are available at http://echa.europa.eu/information-on-chemicals /registered-substances.

[‡] For reviews of the scientific literature on HBCDD, see also Arnot et al. (2009) and US EPA (2014).

Before the enactment of the REACH Regulation in 2006, HBCDD was listed on the second priority list* for the EU's Existing Substances Regulation.[†] The EU's risk assessment of HBCDD therefore began before the enactment of REACH, with the Swedish Chemicals Agency acting as the competent authority performing the analysis. HBCDD has also been subject to a Voluntary Emissions Control Action Programme for brominated flame retardants, which may be responsible for some recent gains in emissions control (VECAP 2013).

The Swedish Chemicals Agency (2007) released a Strategy for Limiting Risks of HBCDD as per the Existing Substances Regulation. The Risk Reduction Strategy includes some risk-related information, but its focus is on reviewing potential legislative tools through which HBCDD might be managed. This strategy reviews potential restrictions under, for example, the Limitations Directive, the Water Framework Directive, the Directive on Hazardous Waste, the Directive on Pollution Prevention and Control, and the Restriction of Hazardous Substances Directive.[‡] The Risk Reduction Strategy also suggests that HBCDD should be included in the Stockholm Convention as a POP. The strategy document precedes the implementation of risk management under the REACH Regulation but it nonetheless demonstrates that other risk management strategies are available in the EU.

In May 2008, the Swedish Chemicals Agency, on behalf of the EU, released its final Risk Assessment Report for HBCDD (Swedish Chemicals Agency 2008). The risk assessment was conducted under the authority of the Existing Substances Regulation using methods specified in the EU's Technical Guidance Document (TGD) on Risk Assessment according to the principles laid out in the Risk Assessment Regulation,[§] which was repealed by REACH. Nonetheless, risk assessment and management of HBCDD under REACH are based on the Risk Assessment Report as well as the Risk Reduction Strategy (ECHA 2008a).

The risk assessment concludes that HBCDD is a PBT and may have toxic effects on reproduction. The PBT assessment was based on the criteria defined in the 2003 TGD. The criteria are somewhat similar to those

---

* Commission Regulation 2268/95, Second List of Priority Substances, 1995 O.J. (L 231) 18 (EC).

[†] Council Directive 793/93, Evaluation and Control of the Risks of Existing Substances, 1993 O.J. (L 84) 1 (EEC).

[‡] Council Directive 76/796, Restrictions on the Marketing and Use of Certain Dangerous Substances, 1976 O.J. (L 262) 201 (EEC); Council Directive 2000/60/EC, Framework for Community Action in the Field of Water Policy, 2000 O.J. (L 327) 1 (EU); Council Directive 91/689, Hazardous Waste, 1991 O.J. (L 377) 20 (EEC); Council Directive 96/61, Pollution Prevention Control, 1996 O.J. (L 257) 26 (EC); Parliament and Council Directive 2011/65, Restriction on the Use of Certain Hazardous Substances, 2011 O.J. (L 174) 88 (EU).

[§] Commission Regulation 1488/94, Assessment of the Risks of Existing Substances, 1994 O.J. (L 161) 3 (EC).

in REACH Annex XIII (European Commission 2003).* One notable difference is that the TGD does not allow for a persistence finding to be based on evidence of a chemical's half-life in soil, while Annex XIII does allow for such a finding. The risk assessment nonetheless bases its persistence finding on evidence that HBCDD has a low degradation rate in soil (a half-life of 210 days according to one study), noting that data are not available to make a definitive persistence finding based on the TGD criteria for half-life in sediment. While one study finds a half-life of 21 and 61 days in two samples, another finds a half-life of over 190 days. The bioaccumulation finding is more conclusive. The risk assessment uses a reported BCF of 18,100 as a representative value, satisfying the criterion to be considered very bioaccumulative (BCF > 5000). The report notes that HBCDD has a ubiquitous presence in the environment. It can be found in remote locations, including the Arctic, and high concentrations have been observed in polar bears, bird eggs, porpoises, and seals. That the chemical can be found in top predators suggests that it accumulates up the food chain. Finally, the report notes several studies on reproductive toxicity, including a reproduction test on *Daphnia magna* that finds a no observed effect concentration of 3.1 µg/L, satisfying the toxicity criteria.

After the release of the risk assessment, administration of assessment and management of HBCDD transitioned into the REACH process. On June 30, 2008, Sweden submitted an Annex XV dossier to ECHA, Member States, and stakeholders to formally identify HBCDD as an SVHC due to its PBT properties. The comment period ended on August 29, 2008. After the Swedish Chemicals Agency edited the dossier, it was referred to the ECHA Member State Committee on September 15, 2008. The committee discussed the dossier at its meeting on October 7 and 8 and, on October 8, 2008, the Member State Committee unanimously agreed to identify HBCDD and its major diastereoisomers as an SVHC due to PBT properties (ECHA 2008b). The SVHC Support Document released with the Member State Committee's decision discusses the evidence presented in the risk assessment (ECHA 2008a). Notably, it affirms the Swedish Chemical Agency's interpretation of the data on persistence and conclusion that HBCDD satisfies the Annex XIII criteria to be designated a PBT.

In accordance with the Member State Committee's opinion, ECHA placed HBCDD on the Candidate List on October 28, 2008, along with

---

* Annex XIII maintains the 2003 Technical Guidance Document's definition of persistence, but adds half-life of >120 days in soil (persistent) and half-life >180 days in soil (very persistent). The 2003 definition of toxicity is broader, allowing a case to be made for chemicals that are category 3 carcinogenic or mutagenic. Annex XIII allows only category 1 and 2. In addition, the 2003 definition of toxicity includes chemicals with endocrine-disrupting effects, while the Annex XIII definition of toxicity does not.

14 other chemicals.* In December 2008, ECHA consulted with its Member State Committee in preparation of a draft recommendation to include HBCDD and six other priority substances on the Annex XIV Authorization List. ECHA published the draft recommendation on January 14, 2009. After reviewing the recommendation, the Member State Committee released its opinion on May 20, 2009, supporting the inclusion of the seven prioritized substances on the Authorization List (ECHA 2009b). The European Commission's REACH Committee approved the addition of the HBCDD to the Authorization List in September 2010, and the European Parliament formally added it to Annex XIV on February 18, 2011.

Inclusion of HBCDD on the Authorization List requires manufacturers, importers, and downstream users to phase out the chemical and replace it with an alternative. The sunset date, after which HBCDD may no longer be used or placed on the market, is August 21, 2015. The deadline for applications for authorization was February 21, 2014. By that deadline, one group had submitted two related applications. To date, ECHA has not made a decision on the applications. ECHA's decision on the HBCDD applications will be the first on whether or not to authorize a use of a PBT listed in Annex XIV.

The applications were submitted by an HBCDD Consortium (2013a) of 13 coapplicants, including joint registrants and downstream users. The first application is for the use of HBCDD as a flame retardant additive in the formulation of unexpanded EPS beads. The second application is for the use of the unexpanded EPS beads to make expanded beads and boards for use in building applications. For the two uses, the consortium submitted a CSR, an Alternatives Analysis, and a Socio-Economic Analysis (HBCDD Consortium 2013a, 2013b, 2013c). The applicants request an expected quantity of 8000 tonnes per year from the sunset date in 2015 through 2019, a four-year "bridging period" or "review period" after which it will be possible to determine whether an alternative can fulfill the demand for a flame retardant additive to EPS in place of HBCDD (HBCDD Consortium 2013a, p. 127).

Notably, the alternatives assessment identifies a newly commercialized alternative substance that is technically feasible for use as a flame retardant additive in the formulation of EPS and is likely not a PBT. The alternative is benzene, ethenyl-, polymer with 1,3-butadiene, brominated, a polymeric flame retardant (pFR) (HBCDD Consortium 2013b). The applicants indicate that, from the samples that have been provided, the pFR will be a technically feasible substitute. The alternative assessment reviews available data on hazards, concluding that it has a low potential

---

* ECHA Decision ED/67/2008, Inclusion of Substances of Very High Concern in the Candidate List, October 28.

for both bioaccumulation and toxicity, but it is intentionally persistent. US EPA (2014) has conducted an alternative assessment of the pFR through the Design for the Environment program and concluded that it is a safer alternative to HBCDD.

The consortium requests a four-year bridging period for five reasons (HBCDD Consortium 2013a). First, they calculate that there will likely not be an adequate supply of the pFR available by HBCDD's sunset date. Second, limitations in supply could generate an "anticompetitive situation" where certain formulators have access to the alternative chemical before others. Third, many Member States have not yet approved the pFR for use in EPS to meet national fire safety standards. Fourth, the market has not yet established confidence in pFR as a feasible alternative. Additional testing is needed on the commercial supply of pFR, and companies throughout the supply chain will need to test the use of pFR in their products. Finally, a ban on HBCDD before there is an adequate supply of the polymeric alternative would have undesirable economic effects, including potential plant closures and resultant loss of jobs.

The authorization applications were opened for comment on May 14, 2014, and the comment period ended on July 9, 2014. Notably, several comments submitted by manufacturers of the polymeric alternative argue that it is already economically feasible and, therefore, ECHA should refuse to grant the authorization application (Chemtura Europe Sales BV 2014). In addition to providing evidence on the availability of the pFR alternative, the commenters take issue with the methodology that the applicants use to calculate the demand for HBCDD for use in EPS formulation, arguing that demand for HBCDD is lower than the applicants project.

As ECHA considered the authorization application, HBCDD was added to Annex A of the Stockholm Convention in November 2014, banning its use, manufacture, and placement on the market worldwide. However, the use of HBCDD in EPS and XPS applications for building materials is exempted from the ban for five years, and the exemption may be extended another five years.* Approval of the authorization request would therefore match the Stockholm Convention exception.

The HBCDD authorization process constitutes a useful test case. It demonstrates how a PBT can move through the authorization process. For HBCDD, the assessment and regulatory decision making processes moved relatively quickly: The risk assessment was released in June 2008 and, by May 2009, ECHA had recommended that it be included on the Authorization List. In this regard, HBCDD may be a bit of an outlier. It did

---

* *Commission Proposal for a Council Decision on the position to be adopted, on behalf of the Europe Union, at the Sixth Conference of the Parties to the Stockholm Convention on Persistent Organic Pollutants with regard to the proposal for an amendment of Annexes A and B,* COM (2013) 134 final (March 12, 2013).

not have to undergo RMO analysis, which would precede the addition of the chemical to the Candidate List. It was also one of the first chemicals that ECHA considered for inclusion on the Candidate and Authorization Lists.

Nonetheless, it is still a useful case study because it is the first PBT on the Authorization List for which ECHA will have to make an authorization application decision. HBCDD might also be unique in that there is a technically feasible alternative that potentially is economically feasible as well. It is possible that ECHA's decision will not turn on HBCDD's PBT properties, but rather on the particularities of the socioeconomic analysis in this case. Even so, ECHA's response to the application should provide some insight into how it will address applications for authorized uses of PBTs in the future. If the decision turns primarily on the economic analysis and the readiness of the alternative, then, perhaps, applicants for PBT and vPvB chemicals will be encouraged to make a socioeconomic case in the future. However, if ECHA rejects the application with the emphasis on HBCDD's PBT properties, then future applicants might decide to forego investing resources into applying for an authorized use of a PBT or vPvB. The point of the authorization mechanism lies in the ability of an applicant to demonstrate that the benefits of a particular use outweigh its risks. If the authorization mechanism becomes a de facto ban on every listed chemical, then it begs the question as to why it is even needed as a risk management tool given that restriction can accomplish that same objective. Authorization might then fail to meet the principle of differentiation of uses and the principle of risk and exposure.

b.   Restriction of phenylmercury as a PBT-like substance of equivalent concern   In 2010, the Norwegian Climate and Pollution Agency (now the Norwegian Environment Agency) submitted an Annex XV Restriction Report to propose a restriction on five phenylmercury compounds: phenylmercury acetate, phenylmercury propionate, phenylmercury 2-ethylhexanoate, phenylmercury octanoate, and phenylmercury neodecanoate. These compounds are generally used in polyurethane in gaskets, seals, electronics, water-resistant coatings and in consumer products such as chairs, roller skates, and shoe soles (ECHA 2011c). In 2012, the commission enacted a restriction based on the report's recommendation. The restriction is essentially a ban: After October 10, 2017, the compounds shall not be manufactured, placed on the market, or used as substances or in mixtures or articles if the concentration of mercury in the mixture or article is equal to or greater than 0.01% by weight.* The phenylmercury compounds are not listed on the Candidate List.

---

* Commission Regulation 848/2012, Amending Annex XVII of REACH as regards Phenylmercury Compounds, 2012 O.J. (L 253) 5 (EU).

The PBT assessment of the compounds themselves shows that phenylmercury meets the Annex XIII toxicity criterion, but it does not meet the criteria for persistence or bioaccumulation. However, the phenylmercury compounds will degrade or transform into methylmercury in the environment. Methylmercury has been determined to be a PBT-like substance. It meets the criteria for toxicity. Recall from Chapter 1 the methylmercury incidents in Minamata Bay, Japan, and with the Huckleby family in New Mexico. Exposure to methylmercury through ingestion has been shown to cause toxic effects to the neurological system. It is also highly bioaccumulative, surpassing the threshold to be considered "very bioaccumulative" (ECHA 2011c).

Persistence, though, is difficult to measure for methylmercury because methylation (degradation of the phenylmercury compounds into methylmercury) and demethylation (degradation of methylmercury into inorganic mercury) occur simultaneously. Persistence must therefore be determined by the ratio of methylation to demethylation. Because of the complexities of this process, persistence cannot be confirmed based on available data. However, data show that methylation occurs at a higher rate than demethylation. In addition, methylmercury has a high biological half-life, with some studies showing that it may persist in some fish species for up to two years and in humans between 44 and 80 days (ECHA 2011c). Authorities have therefore determined that methylmercury is a "PBT like substance or a substance of equivalent concern" to PBTs (ECHA 2011c, pp. 151–152). The Annex XV restriction proposal thus refers to it as a PBT-like substance. If the degradation or transformation products of a substance satisfy the criteria for designation as a PBT, vPvB, or substance of equivalent concern, and the products will be generated in individual amounts equal to or greater than 0.1% of the weight of the original chemical over the course of a year, then the original chemical is treated as a PBT, vPvB, or substance of equivalent concern (ECHA 2014k). Hence, phenylmercury compounds are restricted based on their status as a "substance of equivalent concern" to PBTs.

The restriction was enacted in 2012 and takes effect in 2017, approximately five years from the date of its enactment. The restriction proposal evaluated this option against a similar restriction that would have taken effect in 2014, two years after the date of enactment. Considering differences in effectiveness (in terms of protection of human health and the environment), proportionality, practicality (including implementation and enforcement), and monitoring capacity, the proposal suggested the five-year option to allow industry the lead time to develop safer alternatives and to allow government authorities lead time to enhance their monitoring capabilities. The proposal also performed a socioeconomic analysis of the benefits and costs of the restriction proposals. It concluded that community-wide action was necessary owing to the nature of manufacturing and use patterns as well as mercury's capacity for long-range transport and high hazards to the environment.

The restriction on phenylmercury serves as a useful case example in four regards. First, it demonstrates that, with regard to a PBT (or PBT-like substance), the restriction authority may be used as a ban rather than a management option that targets specific uses, as is common with other categories of chemicals subject to restrictions. Second, this case demonstrates how REACH applies to degradation products that raise PBT concerns. In this case, the commission enacted a restriction on a group of chemical compounds that degrade into a PBT-like substance. However, it is reasonable to envision that the authorization process could apply to a chemical whose degradation product raises PBT concerns as well.

Third, the phenylmercury restriction is an example of the application of the substance of equivalent concern concept. In this case, that concept was used to apply the SVHC classification to a substance that meets the toxicity and bioaccumulation criteria, but not necessarily the persistence criterion as it is defined in Annex XIII. In this case, it was not unreasonable for authorities to judge methylmercury to be persistent under a broader understanding of what persistence entails. Methylmercury degrades at a slower rate than it is created, it has a long biological half-life, and it eventually degrades into elemental or inorganic mercury, which have risks of their own. Nonetheless, this use of the equivalent concern category raises interesting questions about how far the category could be stretched to include substances that are not quite PBTs but rather are PBT-like. Moreover, it raises questions about how the equivalent concern category might be used alongside the weight of evidence approach in PBT determinations.

The Norwegian government submitted its restriction proposal for phenylmercury in 2010, and the weight of evidence concept was amended into REACH Annex XIII in 2011. If sufficient data were available, perhaps the report could have used a weight of evidence approach to conclude that methylmercury satisfies the persistence criteria, classifying it as a PBT rather than a PBT-like substance. Does the equivalent concern category allow authorities a back door around weight of evidence decision making when data are unavailable or unreliable? Although we do not disagree with the application of the equivalent concern category in this instance, we caution European authorities (as noted above) to apply the category with great transparency in the future.

Fourth, from a legal perspective, this regulatory justification raises the question of why it was necessary in the first place to reach a substance of equivalent concern or PBT conclusion given that restrictions can be applied to any chemical, including non-SVHC chemicals, for which risks are considered unacceptable (or not adequately controlled under Article 69). In other words, the commission could have applied a restriction

simply based on exposure information in combination with bioaccumulation and toxicity (partial PBT) concerns without invoking a persistence
finding to classify methylmercury as a PBT-like substance. As we noted
in Chapter 3, though, we are concerned that consideration of partial PBTs
may dilute the meaning and usefulness of the PBT category. Therefore,
consideration of all three criteria is certainly a positive aspect of this
restriction proposal. It nonetheless raises questions regarding the application of restrictions to PBTs and partial PBTs.

c. Deca-BDE and RMO

Bis(pentabromophenyl) ether, known as deca-BDE, is used as an additive
flame retardant primarily in plastics and textiles that are used in vehicles and buildings, for example, in upholstery, furniture, roofing, insulation, and piping (ECHA 2014l). It is also used in adhesives, sealants,
coatings, and inks. Deca-BDE belongs to the group of PBDE congeners.*
Although it is not manufactured in the EU, it is imported as a substance
and in articles. Five companies jointly registered deca-BDE, reporting an
annual importation of 10,000 to 100,000 tonnes, although the importation
volume has been declining (ECHA 2014l).

The proposed management of deca-DBE provides an example of how
the various risk management processes under REACH interact with one
another with regard to PBTs. As noted earlier in this chapter, European
authorities originally envisioned that all SVHCs would be added to the
Candidate List and that all substances on the Candidate List would eventually be added to the Authorization List. However, the *SVHC Roadmap* and
Implementation Plan, which outline the RMO process, suggest that the commission and ECHA have the authority to consider restriction under REACH
or risk management under different legislation after a chemical has been
identified as an SVHC. Additionally, the *SVHC Roadmap* and *Implementation
Plan* seem to envision that substances on the Candidate List could be subjected to restriction rather than authorization, even if the initial presumption
favors the authorization route (ECHA 2013b). Yet, the commission has not
put forward any formal procedure for determining which chemicals and
uses should be a priority for regulation under the restriction approach or
when the restriction approach will generally be preferred to authorization.
The proposed management of deca-BDE provides some insight on the choice
of restriction over authorization for a substance on the Candidate List.

In 2012, ECHA added deca-BDE to the Candidate List as a PBT and
vPvB chemical. The SVHC support document indicates that while deca-
BDE itself satisfies the criteria to be considered "very persistent," it does
not satisfy the criteria for bioaccumulation or toxicity (ECHA 2012).

---

* A detailed description of the uses and risks associated with PBDE flame retardants is
presented as part of a case study in Chapter 6.

However, degradation of deca-BDE in soil produces PBDE congeners with fewer bromine atoms. The congeners tetra-, penta-, hexa-, and hepta-BDE have all been shown to meet the PBT and vPvB criteria. Therefore, authorities designated the deca-BDE parent chemical as a PBT/vPvB as well (ECHA 2012).

Under the supposed early understanding of SVHC identification and placement of a chemical on the Candidate List, it would have seemed that deca-BDE was slated for management under the authorization mechanism. However, in 2013, the Norwegian government submitted a proposal under the EU's POPs Regulation (850/2004/EC), which implements the Stockholm Convention, to include deca-BDE in the convention as a POP. In response, the commission requested that ECHA prepare a complementary Annex XV restriction proposal for deca-BDE. ECHA therefore declined to recommend deca-BDE for inclusion on the Annex XIV Authorization List "to avoid potential regulatory uncertainty" (ECHA 2014l, p. 45).

In August 2014, ECHA (2014l), in collaboration with the Norwegian Environment Agency, submitted an Annex XV Restriction Report to propose a restriction on deca-BDE. The proposed restriction is to prohibit the manufacture, use, and placement on the market of deca-BDE as a substance or as a constituent, in a mixture, or in articles if the concentration in the mixture or article is equal to or greater than 0.1% by weight. The proposal provides for exemptions for articles in use before the restriction's entry into force, for electrical equipment within the scope of the Restriction on Hazardous Substances Directive (2011/65/EU), and for certain aircraft components (ECHA 2014l).

Currently, there is debate and uncertainty as to when authorization or restriction should be the preferred risk management approach under REACH. Although the RMO process is structured to presumptively favor authorization, there is uncertainty regarding how the two risk management approaches interact with one another or when restriction may be chosen over authorization. Recall that authorization applies to the use and placement of a substance on the market, but not to the use of the substance in imported articles (Herbatschek, Bergkamp, and Mihova 2013; Bergkamp and Herbatschek 2014). REACH indicates that if a chemical has been placed on the Authorization List, then ECHA must, after the sunset date, assess "whether the use of the substance in articles poses a risk to human health or the environment that is not adequately controlled."* If ECHA finds that the risk is not adequately controlled, then it must prepare an Annex XV restriction proposal. This seems to suggest that restriction will not be considered for chemicals on the Authorization List until after the sunset date, by which the prohibition takes effect (Herbatschek, Bergkamp, and Mihova 2013). However, REACH Article 58(6) indicates

* REACH, Art. 69(2).

that new restrictions can be placed on chemicals on the Authorization List to address risks that arise from the presence of the chemical in an article. At present, there is little experience to shed light on how restrictions might be placed on Annex XIV chemicals. There is also no guidance for how restrictions might be placed on chemicals that are on the Candidate List but not the Authorization List. The *SVHC Roadmap* and *Implementation Plan* suggest, in theory, that authorities can choose a risk management path for chemicals on the Candidate List other than authorization. The restriction proposal for deca-BDE is an example of this choice in practice.

The restriction proposal for deca-BDE demonstrates an additional point as well. Just as with phenylmercury, the restriction proposal for deca-BDE is essentially a ban on the substance (although with some very limited exceptions). The restriction on polychlorinated terphenyls is also a ban.*

The attractiveness of the restriction mechanism is that it can target particular uses that present risks rather than banning all uses of a chemical. The case of deca-BDE along with phenylmercury and polychlorinated terphenyls (PCTs) together raise the question as to whether restrictions applied to PBTs (and PBT-like substances) will target specific uses or whether they will usually take the form of bans with limited exemptions for particular uses. If the commission is not concerned about the safety of some specific uses, the stigmatization effect of the listing processes under authorization may seem less appropriate than a targeted restriction of the risky uses.

Unacceptable risk (lack of adequate control) in terms of substances with PBT properties therefore reflects the notion that there is no safe level of emissions as well—that any release of a PBT/vPvB to the environment (and subsequent exposure) constitutes a lack of adequate control and, therefore, unacceptable risk.

The tendency to ban PBT and vPvB substances, whether through authorization or restriction, is a reflection of how European authorities view PBT properties. Recall that PBTs, vPvBs, and substances of equivalent concern are considered under REACH to be nonthreshold substances for which there is no safe level of exposure. Unacceptable risk in terms of substances with PBT properties might also therefore reflect the notion that there is no safe level of emissions—that any release of a PBT/vPvB to the environment (and subsequent exposure) constitutes a lack of adequate control and, therefore, unacceptable risk (as per Article 69). The restriction proposal report for deca-BDE illustrates this view, reinforced by the notion that quantitative risk assessment is not possible for PBTs or vPvBs because of the synergistic effects of persistence and bioaccumulation together:

---

* See REACH, Annex XVII and ECHA (2015e).

> Experience with PBT/vPvB substances has shown
> that they give rise to specific concerns based on
> their potential to accumulate in the environment
> and cause effects that are unpredictable in the long-
> term and are difficult to reverse (even when emis-
> sions cease). Therefore, the risk from PBT/vPvB
> substances cannot be adequately addressed in a
> quantitative way, e.g., by derivation of PNECs [pre-
> dicted no effect concentration] and a qualitative risk
> assessment should be carried out…. Emissions and
> subsequent exposure, in the case of a PBT/vPvB, can
> also be usefully considered as a proxy for unaccept-
> able risk (ECHA 2014l, p. 9).

This underlying view of PBTs is shared unevenly around the world
(see Chapter 5). Japan's system seems to reflect a similar view given
that substances with all three PBT properties are subject to prohibition.
However, Japan has given more emphasis to exposure over time, and regulatory authorities in North America do not treat PBTs as nonthreshold
substances.

7.   Risk management principles

Although it is still early in the process, some trends can be discerned with
respect to the key question of whether REACH will be implemented in
ways that are consistent with the management principles. Some of the features of REACH are worrisome, while others are positive. Overall, maintaining consistency with the principles requires a bit of balancing.

In terms of information quality and the value of information, REACH
seems to be heading in the right direction. The purpose of registration is
to protect human health and the environment through filling data gaps
on chemicals in commerce. It seeks to accomplish this by placing the burden of proof of safety and the burden of producing data on industry. The
key to fulfilling these informational management principles is to ensure
that the dossiers include high-quality information while focusing informational investments on data that will be useful to the regulators and
the marketplace, a balancing task that is not easily accomplished. At first
glance, registration seems overly broad (Abelkop and Graham 2015). In
addition, some initial reviews of registration dossiers indicate that information quality may be lower than desired (e.g., Rudén and Hansson 2010;
Gilbert 2011; Rovida, Longo, and Rabbit 2011; Scheringer 2013; Ball et
al. 2014; Stieger et al. 2014). Yet, the purpose of imparting informational
requirements on a broad range of chemicals in commerce is to provide
the marketplace with a baseline of information not only on which chemicals have concerning properties (e.g., PBT, CMR, or endocrine disrupting

properties), but also on which chemicals are safe for their intended uses, thereby improving public confidence in the chemical industry. Baseline informational requirements also help to identify gaps in the data and to focus efforts on determining what investments to make to fill those information gaps. The work of the PBT Expert Group as well as the efforts of ECHA in dossier evaluation and Member States in substance evaluation can help to both identify gaps in registration data and determine what data, if any, need to be generated to allow industry and government to make confident decisions on safety and the desirability of any additional management.

Maintaining consistency with the precautionary principle on the one hand and the principle of exposure and risk assessment and the principle of differentiation of uses on the other seems to also require some balancing with respect to PBTs. These principles are not inherently inconsistent with one another—precautionary decision making can be based in risk rather than hazard. Still, REACH's approach to PBTs, formally addressing them as nonthreshold substances, seems to be unnecessarily unbalanced. The notion that the risks of PBTs cannot be reliably characterized through quantitative risk assessment and that there is no safe exposure level to PBTs is certainly precautionary. However, this approach to PBT regulation is unverified and inconsistent with practices elsewhere in the world, and there is nothing inherent to the precautionary principle that would dictate such an approach.

As of yet, though, the extent to which this approach affects risk management decision making under REACH is unclear. The utility of the restriction mechanism is that it allows regulators to focus management efforts on particular uses that present unacceptable risk. Yet, for PBTs (or at least some PBTs), the restriction mechanism seems to function more like authorization—a ban with exemptions for certain uses. One thing for regulators to consider when using the restriction mechanism this way is that there is no appeal procedure (as in authorization) to allow regulators to exempt uses from the ban after industry generates data to make a positive case for a particular use.* Under authorization, applicants cannot apply for a continued use based on a justification that risks are adequately controlled because the legislation views any exposure level of PBTs as unsafe and hence inadequately controlled. Applicants must therefore make a favorable socioeconomic case for the continued use of a PBT. Allowing applicants to weigh the socioeconomic benefits against the risks of the particular use, though, indirectly allows them to make a case that risks of a particular use are low, in which case even a moderate economic benefit may be enough to favor authorization of the use. ECHA's HBCDD

---

* For a comparative examination of the choice between authorization and restriction as risk management options, see Bergkamp and Herbatschek (2014).

case will therefore be an important test case that sets a precedent for how the agency will address socioeconomic justifications for continued use of a PBT. If the status of PBTs as nonthreshold substances seems to unnecessarily predispose ECHA to reject the application, then the authorization mechanism may be inconsistent with the principle of differentiation of high-risk/low-risk and high-benefit/low-benefit uses.

Overall, considering RMO for all PBTs before placing them on the Candidate List is a positive step and is consistent with the management principles. In addition, REACH is consistent with the rational alternatives and priority-setting principles. Alternatives assessment, including evaluating the risks of alternatives, is integrated into several parts of the REACH Regulation. Before it becomes a basis for risk management decision making, the PBT concept is used to prioritize chemicals for SVHC assessment and RMO analysis. While the complex design of REACH is resource-intensive (Abelkop et al. 2012), we believe that intelligent implementation of REACH can be faithful to the management principles while achieving the legislation's precautionary objectives.

IV.   Conclusion

In this chapter, we have surveyed international and regional chemical policies that utilize the PBT concept. These policies come in a variety of forms. Laws at every level of governance utilize each of the three types of PBT policies that we have identified: PBT properties are used to prioritize chemicals for assessment and regulation, to trigger information-gathering requirements, and to elicit decisions on risk management.

At the international level, SAICM and the Stockholm Convention on Persistent Organic Pollutants take very different approaches to chemical regulation. SAICM is a nonbinding framework that establishes a risk management goal and provides guidance to nations and stakeholders to help them achieve that goal. The WSSD 2020 goal drives policies (including PBT policies) at all levels of governance. The WSSD 2020 goal and SAICM do not apply to any particular chemical or chemical type but do focus some effort on PBTs.

The Stockholm Convention on Persistent Organic Pollutants, on the other hand, focuses exclusively on a subcategory of PBTs that also meet a long-range transport criterion. Unlike the WSSD and SAICM, the Stockholm Convention imparts binding risk management obligations on parties to the convention. Because of its global scope, it is natural for the treaty to limit its focus, as only a subset of chemicals will elicit global concern, and because of the high transaction costs associated with lawmaking at the international level. As the number of sovereign actors involved in the decision-making process increases, cooperation becomes more difficult to sustain (Barrett 2003).

Regional policies also take multiple forms. The GLWQA is a non-binding framework that facilitates cooperation among federal authorities, state and provincial authorities, and nongovernmental stakeholders in the Great Lakes Basin. Previous Great Lakes agreements have utilized the PBT concept to focus risk assessment efforts, and it seems likely that the newest agreement will also utilize PBT properties to prioritize chemicals for assessment and coordinated management.

Like the Great Lakes agreements, the OSPAR Convention focuses on the environmental quality of a particular geographical area. It utilizes the PBT concept not only for prioritization efforts but also for information gathering and risk management. It serves as a useful example of how different methodological approaches to determining what is, and is not, persistent, bioaccumulative, and toxic can alter the number of chemicals that fall into the PBT category.

OSPAR has ceded assessment and management of PBTs and other chemicals to the EU, which regulates existing substances primarily through the REACH Regulation. Although REACH, as EU law, is most accurately characterized as international or regional, it shares some characteristics with national laws. The other regional and international laws that we have discussed require other sovereign governmental units to enact some form of implementation legislation. REACH, on the other hand, imparts legal obligations not exclusively on other units of government (e.g., Member States) but directly on private actors as well, just as national and subnational laws do. Like the laws we discuss in Chapter 5, the REACH Regulation is a vehicle through which authorities may require industry to generate and provide data and may impose risk management directly on the use, manufacture, or importation of chemicals. It is therefore appropriate and useful to compare REACH with chemical legislation at the national level in other jurisdictions.

The international and regional laws are unevenly consistent with the risk management principles. They all seem to fulfill the priory-setting principle by utilizing the PBT concept to focus assessment and regulatory efforts. In addition, each seems to be generally consistent with the principle of rational alternatives. All of the polices emphasize information quality and include features to encourage use of laboratory and field evidence over modeling estimates, although gaps in data remain. Finally, they are all based in the precautionary principle. The PBT concept itself is a precautionary heuristic, and the combination of persistence and bioaccumulation certainly makes risk assessment of a chemical more difficult, although not impossible.

The unpredictable nature of PBTs in the environment may generate an inclination to prefer legal frameworks that pressure for substitutes first and assess uses and releases second, as with REACH authorization (and possibly the restriction mechanism as well). Treating PBTs as nonthreshold substances, though, is not consistent with the principle of exposure and risk assessment. If the socioeconomic route to continue valuable uses of PBTs is shown to be workable, REACH will have sidestepped the allegation that the law automatically bans all uses of chemicals, even high-benefit uses, that fall into the PBT category.

The key is to base decisions on risk, evaluating both exposure and hazard, and to allow for risk management decisions to differentiate between high-risk and low-risk and between high-benefit and low-benefit uses, even for PBTs. This is particularly important given that some of the properties, especially persistence, may be commercially desirable (as with the pFR alternative to HBCDD). Decision-making processes like the RMO are a positive step because they allow for authorities to focus on hazardous chemicals in a way that allows them to evaluate a wider range of policy tools, including efforts to limit releases in particular high-benefit applications, rather than on automatic bans.

In the next chapter, we analyze national PBT policies, comparing them with REACH and other regional and international approaches where appropriate.

References

Abelkop, A.D.K., and J.D. Graham. 2015. "Regulation of Chemical Risks: Lessons for Reform of the Toxic Substances Control Act from Canada and the European Union." *Pace Environmental Law Review* 32.

Abelkop, A.D.K., Á. Botos, L.R. Wise, and J.D. Graham. 2012. "Regulating Industrial Chemicals: Lessons for U.S. Lawmakers from the European Union's REACH Program." Bloomington, IN: Indiana University School of Public & Environmental Affairs. Available at http://www.indiana.edu/~spea/faculty/pdf/REACH_report.pdf.

AGL (Alliance for the Great Lakes), Biodiversity Project, Canadian Environmental Law Association, and Great Lakes United. 2007. *The Great Lakes Water Quality Agreement: Promises to Keep; Challenges to Meet.* Available at http://www.precaution.org/lib/promises_to_keep_challenges_to_meet.061221.pdf.

Arnot, J.A., L. McCarty, J. Armitage, L. Toose-Reid, F. Wania, and I. Cousins. 2009. "An Evaluation of Hexabromocyclododecane (HBCD) for Persistent Organic Pollutant (POP) Properties and the Potential for Adverse Effects in the Environment." Submitted to European Brominated Flame Retardant Industry Panel, May 26.

Ball, N., M. Bartels, R. Budinsky, J. Klapacz, S. Hays, C. Kirman, and G. Patlewicz. 2014. "The Challenge of using Read-Across within the EU REACH Regulatory Framework; How Much Uncertainty is Too Much? Dipropylene Glycol Methyl Ether Acetate, an Exemplary Case Study." *Regulatory Toxicology and Pharmacology* 68:212–21.

Barrett, S. 2003. *Environment & Statecraft: The Strategy of Environmental Treaty-Making.* New York: Oxford University Press.

Bengtsson, G. 2010. "Global Trends in Chemicals Management." In *Regulating Chemical Risks: European and Global Challenges*, edited by J. Eriksson, M. Gilek, and C. Rudén, 179–215. New York: Springer.

Bergkamp, L., ed. 2013. *The European Union REACH Regulation for Chemicals: Law and Practice.* New York: Oxford University Press.

Bergkamp, L., and N. Herbatschek. 2014. "Regulating Chemical Substances under REACH: The Choice between Authorization and Restriction and the Case of Dipolar Aprotic Solvents." *Review of European Community & International Environmental Law* 23:221–45.

Buccini, J. 2004. *The Global Pursuit of the Sound Management of Chemicals.* Washington, DC: World Bank.

Chemtura Europe Sales BV. 2014. "Submission of Information on Alternatives for Applications for Authorisation." Consultation 0013–01 and 0013–02 (HBCDD), June 18. Available at http://echa.europa.eu/documents/10162/18074545/a4a_comment_553_1_attachment_en.pdf.

COP (Conference of the Parties of the Stockholm Convention on Persistent Organic Pollutants). 2009. "Initiation of a Cooperative Framework to Support Parties to Eliminate Polychlorinated Biphenyls through Environmentally Sound Management and Disposal." UNEP/POPS/COP.4/9/Rev.1. Fourth Meeting, Geneva, May 4–8.

CSES (Centre for Strategy and Evaluation Services). 2012. "Interim Evaluation: Impact of the REACH Regulation on the Innovativeness of the EU Chemical Industry," June 14. Available at http://ec.europa.eu/enterprise/sectors/chemicals/files/reach/review2012/innovation-final-report_en.pdf.

DeHihns III, L.A., M. Hey, and C.M. Zygmont. 2009. "Better Living through Green Chemistry." *Natural Resources & Environment* 24:19–23.

Ditz, D.W. 2007. "The States and the World: Twin Levers for Reform of U.S. Federal Law on Toxic Chemicals." *Sustainable Development Law & Policy* 82:27–30.

Ditz, D.W., and B. Tuncak. 2014. "Bridging the Divide between Toxic Risks and Global Chemicals Governance." *Review of European Community & International Environmental Law* 23:181–94.

EC (Environment Canada). 1995. *Toxic Substances Management Policy.* Ottawa, ON, Canada: Environment Canada. Available at http://publications.gc.ca/collections/Collection/En40-499-1-1995E.pdf.

EC (Environment Canada) and US EPA (US Environmental Protection Agency). 1997. "The Great Lakes Binational Toxic Strategy." Available at http://www.epa.gov/greatlakes/p2/bns.html.

EC (Environment Canada) and US EPA (US Environmental Protection Agency). 2009. "Great Lakes Binational Toxics Strategy: 2009 Biennial Progress Report." Available at http://www.epa.gov/bns/reports/2009/2009GLBTSrpt.pdf.

ECHA (European Chemicals Agency). 2007. *Guidance on the Preparation of an Annex XV Dossier for Restrictions,* June. Available at http://echa.europa.eu/documents/10162/13641/restriction_en.pdf.

ECHA (European Chemicals Agency). 2008a. "Member State Committee Support Document for the Identification of Hexabromocyclododecane and All Major Diastereoisomers Identified as a Substance of Very High Concern," October 8. Available at http://echa.europa.eu/documents/10162/13638/svhc_supdoc_hbccd_publication_en.pdf.

ECHA (European Chemicals Agency). 2008b. "Agreement of the Member State Committee on the Identification of Hexabromocyclododecane (HBCDD) and All Major Diastereoisomers Identified as a Substance of Very High Concern According to Articles 57 and 59 of Regulation (EC) No 1907/2006," October 8. Available at http://echa.europa .eu/documents/10162/47d061d9-e336-4139-883b-f00331278cda.

ECHA (European Chemicals Agency). 2009a. "Background Document for Hexabromocyclododecane and All Major Diastereoisomers Identified (HBCDD)," June 1. Available at http://echa.europa.eu/documents /10162/9b8562be-30e9-4017-981b-1976fc1b8b56.

ECHA (European Chemicals Agency). 2009b. "Opinion of the Member State Committee on the Draft Recommendation of the Priority Substances and Annex XIV Entries." ECHA/MSC-8/2009/015, May 20. Available at http:// echa.europa.eu/documents/10162/13576/opinion_draft_recommenda tion_annex_xiv_en.pdf. 182/2011.

ECHA (European Chemicals Agency). 2011a. "Background Document to the Decision of the Executive Director of ECHA, ED/32/2011: Selection Criteria to Prioritise Substances for Substance Evaluation (2011 CoRAP selection criteria)," May 26. Available at http://echa.europa.eu/documents/10162/13628 /background_doc_criteria_ed_32_2011_en.pdf.

ECHA (European Chemicals Agency). 2011b. *Guidance on the Preparation of an Application for Authorisation*, January. Available at http://echa.europa.eu /documents/10162/13637/authorisation_application_en.pdf.

ECHA (European Chemicals Agency). 2011c. "Committee on Risk Assessment and Committee on Socio-Economic Analysis. Background Document to the Opinions on the Annex XV Dossier Proposing Restrictions on Five Phenylmercury Compounds," September 15. Available at http://echa .europa.eu/documents/10162/4a71bea0-31f0-406d-8a85-59e4bf2409da.

ECHA (European Chemicals Agency). 2012. "Member State Committee Support Document for Identification of bis(pentabromophenyl) Ether as a Substance of Very High Concern because of its PBT /vPvB Properties," November 29. Available at http://echa.europa.eu /documents/10162/27064fdb-1cb4-4d37-86c3-42417ec14fb6.

ECHA (European Chemicals Agency). 2013a. "Dossier Evaluation." Document PRO-0017.03, July 8. Available at http://echa.europa.eu /documents/10162/13607/pro_0017_03_dossier_evaluation_en.pdf.

ECHA (European Chemicals Agency). 2013b. *SVHC Roadmap to 2020 Implementation Plan*, December 9. Available at http://echa.europa.eu/documents /10162/19126370/svhc_roadmap_implementation_plan_en.pdf.

ECHA (European Chemicals Agency). 2013c. "Authorizations–Economic Feasibility." Eighteenth Meeting of the Committee for Socio-Economic Analysis. Document SEAC/18/2013/03. Available at http://echa.europa .eu/documents/10162/13580/seac_authorisations_economic_feasibility _evaluation_en.pdf.

ECHA (European Chemicals Agency). 2013d. *Estimating the Abatement Costs of Hazardous Chemicals: A Review of the Results of Six Case Studies*. ECHA-13-R-06-EN, September. Helsinki: ECHA. Available at http://echa.europa.eu /documents/10162/13580/abatement+costs_report_2013_en.pdf.

ECHA (European Chemicals Agency). 2014a. "Registered Substances." Last modified December 19. Available at http://echa.europa.eu /information-on-chemicals/registered-substances.

ECHA (European Chemicals Agency). 2014b. "Substance Evaluation." Document PRO-0023.02, October 20. Available at http://echa.europa.eu /documents/10162/13607/pro_0023_01_substance_evaluation_en.pdf.

ECHA (European Chemicals Agency). 2014c. *Guidance for the Preparation of an Annex XV Dossier for the Identification of Substances of Very High Concern, Version 2.0*, February. Available at http://echa.europa.eu/documents/10162/13638 /svhc_en.pdf.

ECHA (European Chemicals Agency). 2014d. "Prioritisation and Annex XIV Recommendation." Document PRO-0034.02, November 21. Available at http://echa.europa.eu/documents/10162/13607/prioritisation_annex _xiv_recommendation_en.pdf.

ECHA (European Chemicals Agency). 2014e. "Candidate List of Substances of Very High Concern for Authorisation." Last modified December 17. Available at http://echa.europa.eu/candidate-list-table.

ECHA (European Chemicals Agency). 2014f. "Draft Background Document for Coal Tar Pitch, High Temperature," September 1. Available at http://echa .europa.eu/documents/10162/51ba0da7-f342-449d-a3a2-4d926fc05bf0.

ECHA (European Chemicals Agency). 2014g. "Draft Background Document for Anthracene Oil," September 1. Available at http://echa.europa.eu /documents/10162/154645b5-fb85-4068-a88e-acd87aed98ea.

ECHA (European Chemicals Agency). 2014h. "Authorisation to Use a Substance of Very High Concern—First Opinions Adopted." *Press Release*, January 3. Available at http://echa.europa.eu/view-article/-/journal_content/title /authorisation-to-use-a-substance-of-very-high-concern-first-opinions -adopted.

ECHA (European Chemicals Agency). 2014i. "Statistics on Received Applications." Last modified December 11. Available at http://echa .europa.eu/web/guest/addressing-chemicals-of-concern/authorisation /applications-for-authorisation/received-applications.

ECHA (European Chemicals Agency). 2014j. *Prioritisation of Substances of Very High Concern (SVHCs) for Inclusion in the Authorisation List (Annex XIV)*, February 10. Available at http://echa.europa.eu/documents/10162/13640 /gen_approach_svhc_prior_in_recommendations_en.pdf.

ECHA (European Chemicals Agency). 2014k. *Guidance on Information Requirements and Chemical Safety Assessment, Chapter R.11: PBT/vPvB Assessment, Version 2.0*, November. Available at http://echa.europa.eu/documents/10162/13632 /information_requirements_r11_en.pdf.

ECHA (European Chemicals Agency). 2014l. "Annex XV Restriction Report, Proposal for a Restriction, Bis(pentabromophenyl) Ether," August 1. Prepared in Collaboration with the Norwegian Environment Agency. Available at http://echa.europa.eu/documents/10162/a3f810b8-511d-4fd0 -8d78-8a8a7ea363bc.

ECHA (European Chemicals Agency). 2015a. "Management of PBT/vPvB Substances under REACH." Accessed January 8. Available at http://echa .europa.eu/addressing-chemicals-of-concern/substances-of-potential-con cern/pbts-and-vpvbs/management-of-pbt-vpvb-substances.

ECHA (European Chemicals Agency). 2015b. "Role of the Member State Committee in the Authorisation Process." Accessed January 8. Available at http://echa .europa.eu/about/organisation/committees/msc/msc_process_en.asp.

ECHA (European Chemicals Agency). 2015c. "Summary of Obligations Resulting from Inclusion in the Candidate List of Substances of Very High Concern for Authorisation." Available at http://echa.europa.eu /candidate-list-obligations.

ECHA (European Chemicals Agency). 2015d. "Authorisation List." Available at http:// echa.europa.eu/web/guest/addressing-chemicals-of-concern/authorisation /recommendation-for-inclusion-in-the-authorisation-list/authorisation-list.

ECHA (European Chemicals Agency). 2015e. "List of Restrictions." Accessed January 8. Available at http://echa.europa.eu/addressing-chemicals-of -concern/restrictions/list-of-restrictions.

ECHA (European Chemicals Agency). 2015f. "PBT Expert Group." Accessed January 8. Available at http://echa.europa.eu/addressing-chemicals -of-concern/substances-of-potential-concern/pbt-expert-group.

ECHA (European Chemicals Agency). 2015g. "Overview of the Substance Specific Work of the PBT Expert Group." Accessed January 8. Available at http://echa .europa.eu/addressing-chemicals-of-concern/substances-of-potential-concern /pbts-and-vpvbs/echas-pbt-expert-group/overview-outcome-case-discus sions/substance-work-status.

EEB (European Environmental Bureau) and ClientEarth. 2012. *Identifying the Bottlenecks in REACH Implementation: The Role of ECHA in REACH's Failing Implementation.* Brussels: European Environmental Bureau.

European Commission. 2003. "Part II Environmental Risk Assessment." In *Technical Guidance Document on Risk Assessment.* Joint Research Centre, Institute for Health and Consumer Protection, European Chemicals Bureau. Ispra, IT: European Commission.

European Commission. 2013. *Roadmap on Substances of Very High Concern.* 5867/13. Brussels: European Commission, February 6. Available at http://register .consilium.europa.eu/doc/srv?l=EN&f=ST%205867%202013%20INIT.

European Commission. 2014. "Restrictions." DG Enterprise and Industry. Last modified November 21. Available at http://ec.europa.eu/enterprise /sectors/chemicals/reach/restrictions/index_en.htm.

Galatone, V. 2014. "Canada's Chemicals Management Plan and the Great Lakes Water Quality Agreement Annex 3—Chemicals of Mutual Concern." Presentation at the Great Lakes Legislative Caucus Annual Meeting, Québec City, Québec, July 24–25, 2014. Available at http://www.greatlakeslegisla tors.org/LinkClick.aspx?fileticket=dIZG5oc4tz4%3D&tabid=75.

Gilbert, N. 2011. "Data Gaps Threaten Chemical Safety Law." *Nature* 475:150–1.

Godduhn, A., and L.K. Duffy. 2003. "Multi-Generation Health Risks of Persistent Organic Pollution in the Far North: Use of the Precautionary Approach in the Stockholm Convention." *Environmental Science & Policy* 6:341–53.

Grunwald, G., and P. Hennig. 2014. "Impacts of the REACH Candidate List of Substances Subject to Authorisation: The Reputation Mechanism and Empirical Results on Behavioral Adaptations of German Supply Chain Actors." *Journal of Business Chemistry* 11:53–66.

HBCDD Consortium. 2013a. "Socio-Economic Analysis." Available at http://echa .europa.eu/documents/10162/0ed074ab-00e9-42d8-aaea-60a08568551c.

HBCDD Consortium. 2013b. "Analysis of Alternatives: HBCDD Use in EPS for Building Applications." Available at http://echa.europa.eu /documents/10162/0e586519-33dd-41d6-8624-0aee29327e3a.

HBCDD Consortium. 2013c. "Chemical Safety Report—Hexabromocyclododecane." Available at http://echa.europa.eu/documents/10162/ab191f7e-a290-4d75 -b253-da14ce3dd076.

Heitman, K., and A. Reihlen. 2007. "Techno-Economic Support on REACH: Case study on 'Announcement Effect' in the Market Related to the Candidate List of Substances Subject to Authorization." Hamburg, Germany: Ökopol GmbH, Institute for Environmental Strategies. Available at http://ec.europa .eu/environment/chemicals/reach/pdf/background/report_announce ment_effect.pdf.

Herbatschek, N., L. Bergkamp, and M. Mihova. 2013. "The REACH Programmes and Procedures." In *The European Union REACH Regulation for Chemicals: Law and Practice*, edited by L. Bergkamp, 82–170. New York: Oxford University Press.

IOM Consulting. 2009. "Data on Manufacture, Import, Export, Uses, and Releases of HBCDD as well as Information on Potential Alternatives to Its Use." Report prepared for European Chemicals Agency. Available at http://echa .europa.eu/documents/10162/13640/tech_rep_hbcdd_en.pdf.

Karlsson, M. 2010. "The Precautionary Principle in EU and US Chemicals Policy: A Comparison of Industrial Chemicals Legislation." In *Regulating Chemical Risks: European and Global Challenges*, edited by J. Eriksson, M. Gilek, and C. Rudén, 239–65. New York: Springer.

Lipnick, R.L., and D.C.G. Muir. 2000. "History of Persistent, Bioaccumulative, and Toxic Chemicals." In *Persistent, Bioaccumulative, and Toxic Chemicals I: Fate and Exposure*, edited by R.L. Lipnick, J.L.M. Hermens, K. Jones, and D.C.G. Muir. Washington, DC: American Chemical Society.

Maguire, S., and J. Ellis. 2005. "Redistributing the Burden of Scientific Uncertainty: Implications of the Precautionary Principle for State and Nonstate Actors." *Global Governance* 11:505–26.

Moermond, C.T., M.P. Janssen, J.A. de Knecht, M.H. Montforts, W.J. Peijnenburg, P.G. Zweers, and D.T. Sijm. 2012. "PBT Assessment Using the Revised Annex XIII of REACH: A Comparison with Other Regulatory Frameworks." *Integrated Environmental Assessment and Management* 8:359–71.

OSPAR Commission. 2006. *Dynamic Selection and Prioritisation Mechanism for Hazardous Substances (New DYNAMEC Manual)*. London, UK: OSPAR Commission. Available at http://www.ospar.org/documents/dbase/publi cations/p00256/p00256_new%20dynamec%20manual.pdf.

OSPAR Commission. 2010a. "The North-East Atlantic Environment Strategy: Strategy of the OSPAR Commission for the Protection of the Marine Environment of the North-East Atlantic 2010–2020." Agreement 2010-3.

OSPAR Commission. 2010b. "5 Hazardous Substances." In *Quality Status Report 2010*. London: OSPAR Commission. Available at http://qsr2010.ospar.org /en/media/chapter_pdf/QSR_Ch05_EN.pdf.

OSPAR Commission. 2015a. "Chemicals." Accessed January 8. Available at http:// www.ospar.org/content/content.asp?menu=01460304880000_000000_000000.

OSPAR Commission. 2015b. "Decisions, Recommendations, Agreements, Imple-mentation Reports and Formats." Accessed January 8. Available at http:// www.ospar.org/content/dra.asp?menu=01070304570000_000000_000000.

Rovida, C., F. Longo, and R.R. Rabbit. 2011. "How Are Reproductive Toxicity and Developmental Toxicity Addressed in REACH Dossiers?" *ALTEX* 28:273–94.

RTG (Rowan Technology Group). 2015. "REACH." Accessed January 8. Available at http://www.rowantechnology.com/US-and-European-rules/european-regulations/reach.

Rudén, C., and S.O. Hansson. 2010. "Registration, Evaluation, and Authorization of Chemicals (REACH) Is But the First Step—How Far Will It Take Us? Six Further Steps to Improve the European Chemicals Legislation." *Environmental Health Perspectives* 118:6–10.

Scheringer, M. "PBT Assessment." Presentation at the Indiana University School of Public and Environmental Affairs Workshop on PBT Science and Policy, Brussels, Belgium, December 4, 2013.

Selin, H. 2010. *Global Governance of Hazardous Chemicals: Challenges of Multilevel Management*. Cambridge, MA: MIT Press.

Stieger, G., M. Scheringer, C.A. Ng, and K. Hungerbühler. 2014. "Assessing the Persistence, Bioaccumulation Potential and Toxicity of Brominated Flame Retardants: Data Availability and Quality for 36 Alternative Brominated Flame Retardants." *Chemosphere* 116:118–23.

Swedish Chemicals Agency. 2007. "Strategy for Limiting Risks: Hexabromo-cyclododecane (HBCDD). 2nd Priority List," September 4. Available at http://www.dioksyny.pl/wp-content/uploads/RRS-HBCDD-20070904.pdf.

Swedish Chemicals Agency. 2008. "Risk Assessment: Hexabromocyclododecane." Final Report, May. Available at http://echa.europa.eu/documents/10162/661bff17-dc0a-4475-9758-40bdd6198f82.

Swedish Chemicals Agency. 2013. "Stockholm Convention (POPs)." Last modified October 30. Available at https://www.kemi.se/en/Content/International/Conventions-and-agreements/-Stockholm-Convention-POPs/.

Tuncak, B. 2013. *Driving Innovation: How Stronger Laws Help Bring Safer Chemicals to Market*. Washington, DC: Center for International Environmental Law. Available at http://www.ciel.org/Publications/Innovation_Chemical_Feb2013.pdf.

UN (United Nations). 2002. "Plan of Implementation of the World Summit on Sustainable Development." In *Report of the World Summit on Sustainable Development, Johannesburg, South Africa*, August 26–September 4, 2002, A/CONF.199/20. New York: United Nations.

UNEP (United Nations Environment Programme). 2006. *Strategic Approach to International Chemicals Management: SAICM Texts and Resolutions of the International Conference on Chemicals Management*. International Conference on Chemicals Management, Dubai, United Arab Emirates, February 4–6, 2006. Geneva: UNEP. Available at http://sustainabledevelopment.un.org/content/documents/SAICM_publication_ENG.pdf.

UNEP (United Nations Environment Programme). 2014a. "Status of Ratifications." Accessed December 22. Available at http://chm.pops.int/Countries/StatusofRatifications/PartiesandSignatories/tabid/252/Default.aspx.

UNEP (United Nations Environment Programme). 2014b. "What are POPs?" Accessed December 22. Available at http://chm.pops.int/TheConvention/ThePOPs/tabid/673/Default.aspx.

UNEP (United Nations Environment Programme). 2014c. "Chemicals Proposed for Listing Under the Convention." Accessed December 22. Available at http://chm.pops.int/TheConvention/ThePOPs/ChemicalsProposedforListing/tabid/2510/Default.aspx.

UNEP Governing Council. 2002. "Decision SS. VII/3. Strategic Approach to International Chemicals Management," February 15. Available at http://www.chem.unep.ch/irptc/strategy/ss_vii_3.pdf.

US EPA. 2010. "Hexabromocyclododecane (HBCD) Action Plan," August 18. Available at http://www.epa.gov/oppt/existingchemicals/pubs/action plans/RIN2070-AZ10_HBCD%20action%20plan_Final_2010-08-09.pdf.

US EPA. 2014. "Flame Retardant Alternatives for Hexabromocyclododecane (HBCD)." Final Report. EPA Publication 740R14001, June. Available at http://www.epa.gov/dfe/pubs/projects/hbcd/hbcd-full-report-508.pdf.

van Wijk, D., R. Chénier, T. Henry, M.D. Hernando, and C. Schulte. 2009. "Integrated Approach to PBT and POP Prioritization and Risk Assessment." *Integrated Environmental Assessment and Management* 5(4):697–711.

VECAP (Voluntary Emissions Control Action Program). 2013. "Sound Results from a Proactive Industry: European Annual Progress Report 2013." Brussels, Belgium: European Flame Retardants Association and the Bromine Science and Environmental Forum.

Victor, D.G. 1997. "The Use and Effectiveness of Nonbinding Instruments in the Management of Complex International Environmental Problems." *American Society of International Law Proceedings* 91:241–50.

Westervelt, A. 2013. "Exclusive: New Research Links Chemical Regulation with Market Innovation." *Forbes*, February 13. Available at http://www.forbes.com/sites/amywestervelt/2013/02/13/exclusive-new-research-links-chemical-regulation-with-market-innovation/.

Williams, T. 2006. *Virtual Elimination of Pollution from Toxic Substances*. Ottawa, ON, PRB 06-26E: Canada. Canadian Parliamentary Information and Research Service. Available at http://www.parl.gc.ca/Content/LOP/researchpub lications/prb0626-e.htm.

chapter five

National PBT policies in Asia and North America

I.   Introduction

Whereas Chapter 4 considers PBT policies at the international and regional governance levels, this chapter surveys PBT policies at the national level. Although individual European countries do have their own chemical regulations, the European Union (EU) Registration, Evaluation, Authorization, and Restriction of Chemicals (REACH) regulation constitutes the overarching policy through which the laws of Member States function. Therefore, we do not consider the policies of individual European countries. We do consider the national policies of Japan, China, Canada, and the United States.

Each country's PBT policies offer illustrations of how PBT characteristics are incorporated into chemical risk assessment and management regimes. Japan's Chemical Substances Control Law (CSCL) was the first PBT policy in the world, serving as the origin of the PBT concept in a regulatory context. China has become the world's largest chemical manufacturer, but its laws governing chemicals are at an early stage of development and implementation. The relatively streamlined risk assessment and management processes under the Canadian Environmental Protection Act (CEPA) of 1999 and the Chemical Management Plan (CMP) offer a particularly stark contrast to the sheer complexity of REACH. Finally, the United States has incorporated the PBT concept into several of its national regulatory programs, including the Toxic Substances Control Act (TSCA) of 1976, which is primed for revision. The following sections consider each of these laws in turn, drawing comparative insights based on the policy typology and the risk management principles outlined in Chapter 1.

II.   Japan

A.   Chemical Substances Control Law of 1973

Japan's 1973 legislation on industrial chemicals, the Act on the Evaluation of Chemical Substances and Regulation of their Manufacture,* known as

* Kashinho [Chemical Substances Control Law], Law No. 117 of 1973 (hereinafter CSCL).

Kashinho, was the first regulatory program in the world to employ the PBT concept (Toda 2007; Naiki 2010; Uyesato et al. 2013; Ministry of the Economy, Trade, and Industry [METI] 2015). In English, the law is commonly referred to as the Chemical Substances Control Law (e.g., Toda 2007). With the CSCL, the Japanese government established a regulatory framework for evaluating and managing the risks of existing and new chemicals—the first legislation of its kind, preceding similar legislation in Europe (e.g., the 1976 Limitations Directive) and the United States (e.g., the TSCA of 1976). We described the origin of the CSCL in Chapter 1. In this chapter, we discuss the law's substantive use of the PBT concept.

While the amended CSCL employs a mix of hazard- and risk-based approaches to chemical management, the original law was limited in its scope. The Kanemi Yusho disease in the late 1960s, caused by polychlorinated biphenyl (PCB)-contaminated rice oil, focused Japanese regulatory efforts on chemicals that exhibit all three properties: persistence, bioaccumulation, and toxicity. Risk management under the CSCL began with a ban on the manufacture, importation, and use of PCBs, followed by other chemicals that exhibit PBT properties, including most of the chemicals that would later become known as the dirty dozen persistent organic pollutants (POPs) (Shibata and Takasuga 2007; Toda 2007). Under the CSCL, once a chemical is identified as a PBT, it is essentially banned. As in many other jurisdictions, Japan does not have a single comprehensive chemical regulation. The Japanese government focused the CSCL on PBTs, while a variety of other laws already in existence addressed a range of other classes of chemicals and exposure pathways, including agricultural chemicals, pharmaceuticals, cosmetics, occupational exposures, chemicals in household products, and food (Toda 2007; Uyesato et al. 2013).*

In addition to its focus on PBTs, the 1973 CSCL was limited in three other notable ways. First, the 1973 CSCL was entirely hazard based; it did not compel authorities to consider exposure in assessment or management processes. Second, the 1973 CSCL considered effects on human health, but not effects on the environment. Third, the CSCL applied to chemicals that have chronic, long-term toxic effects if ingested continuously, but not to chemicals that have only acute toxic effects (Uyesato et al. 2013). Amendments to the CSCL have eliminated the first two limitations, but the amended CSCL remains focused on long-term toxicity rather than acute toxicity.

---

* Agricultural Chemicals Regulation Act, Law No. 82 of 1942; Pharmaceutical Affairs Act, Law No. 145 of 1960; Industrial Health and Safety Act, Law No. 57 of 1972; Control of Household Products Containing Harmful Substances Act, Law No. 112 of 1973; and Food Sanitation Act, Law No. 233 of 1947.

B.   Amendments to the CSCL

Since its enactment in 1973, the Japanese Diet has expanded the scope of the CSCL through three major legislative amendments. In 1986, the scope of the law was expanded to address chemicals that are found to exhibit persistence and chronic toxicity, but not bioaccumulation (e.g., trichloro-ethylene and tetrachloroethylene) (Toda 2007; Nanimoto 2012). In 2003, in response to a review of its regulatory system, Japan expanded the CSCL again to assess and address ecological effects, to consider limited exposure information, and to identify and manage chemicals that exhibit persistence and bioaccumulation but indeterminate long-term toxicity. The 2003 amendment also increased hazard information reporting requirements on industry (Berger and Marrapese 2004; Toda 2007). Thus, before 2009, the CSCL remained mostly hazard based and applied management restrictions and informational requirements to a variety of partial PBT categories, including chemicals that are persistent and toxic or persistent and bioaccumulative, in addition to full PBTs.

The CSCL is jointly administered by the METI, the Ministry of the Environment, and the Ministry of Health, Labor, and Welfare. Motivated by the World Summit on Sustainable Development (WSSD) 2020 goal and the Strategic Approach to International Chemicals Management, regulatory efforts under the Stockholm Convention, Europe's experience with REACH, and Canada's experience with its CMP, a council of representatives from the ministries authored a report in 2008 proposing additional risk-based modifications to the CSCL (Joint Committee 2008; Government of Japan 2009; METI et al. 2009). In 2009, the Japanese Diet enacted many of those proposals into law, updating the CSCL to reflect a more risk-based approach to assessment and management (Naiki 2010; Nanimoto 2012; Kogan 2013; Uyesato et al. 2013).

PBTs remain a focus of the amended CSCL. Just as with prior laws, PBs are identified and assessed, and PBTs are essentially banned. However, the amended CSCL expanded the PT category to identify any high-exposure-level chemicals that exhibit long-term toxic effects on human health, flora, or fauna, including endocrine disrupting chemicals. In addition, the amended CSCL allows for continued "essential uses" of regulated chemicals, including PBTs (*Chemical Watch* 2011b; METI 2011). In particular, the Japanese government sought to expedite its relatively slow risk assessment process by introducing risk-based prioritization and assessment to identify and manage chemicals of concern (Naiki 2010).

Under the amended CSCL, industry must submit yearly data to the government on existing and new substances. The government screens those data to identify potential PBTs and priority chemicals based on certain hazard and exposure criteria, then engages in stepwise risk assessment and, if necessary, risk management. The process is depicted in Figure 5.1 and described in greater detail in the following sections.

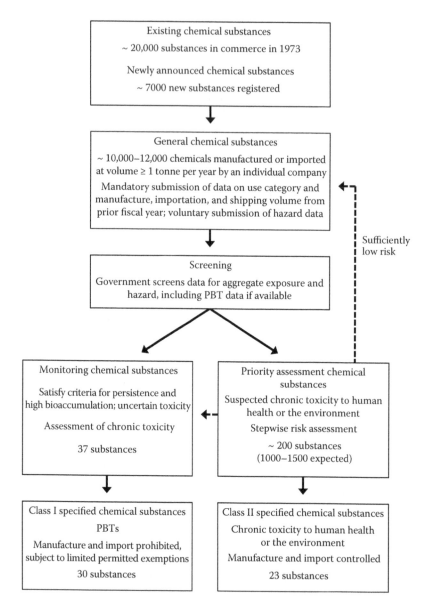

Figure 5.1 Amended CSCL. (Adapted from METI [Ministry of Economy, Trade, and Industry], "Chemical Substances Control Law." Tokyo, Japan: Chemical Safety Office, Chemical Management Policy Division, Manufacturing Industries Bureau, Ministry of Economy, Trade and Industry. Available at http://www .meti.go.jp/policy/chemical_management/english/cscl/files/about/about _points.pdf, 2011 and Naiki, Y., Journal of Environmental Law 22:171–95, 2010.)

C.   Screening of chemicals for PBT properties and prioritization for risk assessment

Before 2009, the CSCL was primarily hazard based by applying risk management to chemicals through PBT and partial PBT classifications. The amended CSCL maintains those hazard-based classifications but instructs authorities to use both exposure and hazard data to prioritize chemicals for risk assessment to determine which chemicals should be managed (METI 2010, 2011). Because the government could not assess all of the existing chemicals by 2020, it established with the amended CSCL a prioritization and tiered risk assessment framework to achieve the WSSD goal (Naiki 2010; Nanimoto 2012; Uyesato et al. 2013).

The regulatory process under the amended CSCL begins with the identification of potential PBTs and priority chemicals for risk assessment. The regulatory scheme for existing chemicals applies to the CSCL's inventory of existing chemicals—approximately 20,600 chemicals that were in commerce in 1973—in addition to about 7000 new chemicals that have been newly announced* to the government before April 2011 (METI 2011; Nanimoto 2012; Uyesato et al. 2013). For these existing and newly announced chemicals, the CSCL requires a company to report on any chemical that it manufactures or imports at a quantity of 1 tonne or more per year.[†] This reporting requirement applies to a constituent chemical in a mixture if the chemical constitutes 10% or more of the mixture by weight (Shi 2013). These chemicals are identified as "General Chemical Substances."[‡]

For every General Chemical Substance that a company imports or manufactures at a volume of 1 tonne or more per year, the CSCL requires the company to submit data to the government on the chemical's identity, use category, shipped volume, and the manufacture and importation volumes from the prior fiscal year (Fukushima 2012; Nanimoto 2012; Shi 2013). Submission of hazard data for General Chemical Substances is voluntary. In 2011, the ministries received nearly 31,000 reports from about 1400 companies on 10,792 chemicals (*Chemical Watch* 2012; Nanimoto 2012; Fukushima 2012; Kimura 2013). In 2012, the government received data on 11,979 chemicals that met the reporting requirements for General Chemical Substances that are then subject to screening assessment (Kimura 2013; Takahashi 2014).[§]

---

* Pre-Amendment CSCL, Art. 4(4). Various documents refer to these chemicals as "newly announced" (METI 2011), "evaluated" (Fukushima 2012), and "notified" (Shi 2013) new chemicals.
[†] CSCL, Art. 8.
[‡] CSCL, Art. 2(7).
[§] Chemicals identified as falling into a separate regulated category are not identified as General Chemical Substances. CSCL, Art. 2(7).

The ministries use the data reported by industry and any other available data to screen the General Chemical Substances. If the available data indicate that the chemical is persistent and highly bioaccumulative,* then the chemical is identified as a "Monitoring Chemical Substance"—a category that designates candidates for identification as "Class I Specified Chemical Substances," which are PBTs.[†] Relatively few chemicals are identified as PBTs and potential PBTs: There are currently 30 Class I PBTs and 37 Monitoring PB chemicals (National Institute of Technology and Evaluation).[‡] All other chemicals are screened to identify priority chemicals for risk assessment, called "Priority Assessment Chemical Substances" (PACSs).[§]

Screening assessment occurs in three steps: determination of an exposure class, determination of a hazard class, and identification of high, medium, and low priorities based on a matrix that considers both exposure and hazard classes (Shimizu 2011; Fukushima 2012; Hirai 2012; Nanimoto 2012; Kimura 2013). The prioritization matrix is depicted in Table 5.1.

To determine a chemical's exposure class, the ministries first consider the data submitted by industry on use category and manufacture and importation volume. Using chemical identification numbers (CAS number or Ministry of International Trade and Industry number), the ministries aggregate the data submitted by multiple companies for a particular chemical. Those chemicals that are, in the aggregate, manufactured and/or imported at a volume of 10 tonnes or more per year are subject to screening assessment (Shimizu 2011; Fukushima 2012; Hirai 2012; Kimura 2013). Substances manufactured and/or imported below the 10 tonnes per year threshold are excluded from further screening for prioritization and remain classified as General Chemical Substances, subject to yearly reporting requirements. Of the 11,979 chemicals that industry reported on, 7819 of them met the 10 tonnes per year threshold and have been subject to further screening (Kimura 2013; Takahashi 2014). For these chemicals, the ministries consider aggregate manufacture and importation volume, usage categories, and an emission factor to estimate total yearly national emissions. Biodegradability information, if available, is then also considered to determine total yearly environmental releases of

---

* For a description of the PBT determination criteria under the CSCL, see Table 2.1. For a description of how persistence, bioaccumulation, and toxicity are determined under the CSCL, see Toda (2007), Nakai, Takano, and Saito (2006), and Hashizume et al. (2013).

[†] CSCL, Art. 2(3) and (4).

[‡] For the latest information on chemical classifications and listings under the CSCL, Japan's National Institute of Technology and Evaluation maintains the Chemical Risk Information Platform at http://www.safe.nite.go.jp/english/sougou/view/SelectingListsList_en.faces and the Japan Chemicals Collaborative Knowledge database at http://www.safe.nite .go.jp/jcheck/top.action. Both websites offer English translations.

[§] CSCL, Art. 2(5). For relevant regulations on PACSs, see CSCL, Art. 9–12.

Table 5.1 Screening scores under the amended CSCL

Exposure class estimated total national emissions (tonnes/year)	Hazard class based on highest score for human health or ecological toxicity			
	Class 1	Class 2	Class 3	Class 4
Class 1 ≥ 10,000	High	High	High	High
Class 2 ≥ 1000	High	High	High	Medium
Class 3 ≥ 100	High	High	Medium	Medium
Class 4 ≥ 10	High	Medium	Medium	Low
Class 5 ≥ 1	Medium	Medium	Low	Low

Source: Adapted from Fukushima, T., "Chemical Legislation and Regulations Updates: Japan." Director for Chemical Management Policy, Chemical Management Policy Division, Ministry of Economy, Trade, and Industry, Japan. Presentation at the Asia-Pacific Economic Cooperation Regulators' Forum, Singapore, March 30, 2012. Available at http://www.estis.net/includes/file.asp?site=cien-peru&file=82080D68-A181-4AFC-B4D0-6B94EBEB46AD, 2012 and Hirai, Y., "Chemical Risk Assessment under the Chemical Substances Control Law in Japan and comparison with REACH." Risk Analysis Division, Chemical Management Center, National Institute of Technology and Evaluation, Japan. Presentation at the Society for Environmental Toxicology and Chemistry: Sixth World Congress/Twenty-Second Europe Annual Meeting, Berlin, Germany, May 20–24, 2012. Available at http://www.safe.nite.go.jp/risk/archive_pdf/SETAC2012_special session_nite.pdf, 2012.

the chemical, which will determine which exposure class the chemical is placed into (Kimura 2013). There are five exposure classes as shown in Table 5.1, ranging from 1 tonne of emissions per year to 10,000 tonnes per year; chemicals in Class 1 (emissions ≥ 10,000 tonnes per year) raise the highest level of concern.

To determine a chemical's hazard class, the ministries assess available hazard information on toxicological endpoints for human health and the environment (*Chemical Watch* 2011a; Shimizu 2011; Fukushima 2012; Hirai 2012; Nanimoto 2012; Kimura 2013). Specifically, the ministries evaluate existing data (e.g., Globally Harmonized System [GHS] classification data) on carcinogenicity, mutagenicity, reproductive toxicity, and repeated dose toxicity for human health. To screen for ecological toxicity, the ministries use available data on a chemical's predicted no-effect concentration. Industry may voluntarily submit hazard data to the government to aid in screening assessments, but it is not mandatory. The ministries determine a hazard class for each toxicological endpoint for a particular chemical among four classes, with Class 1 denoting the highest level of concern and Class 4 denoting the lowest level of concern. Among all of the toxicological endpoints, the highest class will be used as the chemical's overall hazard class for purposes of prioritization. That is, if a chemical is designated

Class 2 for repeated dose toxicity and Class 4 for all other endpoints, the chemical is designated as a Class 2 hazard for prioritization.

The unavailability of data does not denote that a chemical is not hazardous. Rather, insufficient data for repeated dose toxicity and mutagenicity result in a default designation as Class 2. Insufficient data for ecological toxicity will usually result in a default designation of Class 1 (Shimizu 2011). Insufficient data for carcinogenicity and reproductive toxicity will not result in a default classification.

Finally, authorities assess the combined exposure and hazard classes in a matrix (as shown in Table 5.1) and use the matrix to assign chemicals into high-, medium-, and low-priority categories. The ministries classify high-priority substances as PACSs. Medium- and low-priority chemicals may be subjected to further evaluation and either classified as PACSs or remain designated as General Chemical Substances.

Authorities have stated that they expect to eventually designate 1000–1500 chemicals as PACSs (Uyesato et al. 2013). As of October 2014, 164 chemicals have been designated as PACSs (Takahashi 2014). PACSs are subject to additional data reporting requirements and stepwise risk assessment to determine if risk management is appropriate. Companies that manufacture and import PACSs at a volume of 1 tonne or more per year must submit yearly volume and detailed usage information to METI.*

Risk assessment is conducted in stages whereby the government first evaluates primarily exposure-related information, including screening data on yearly aggregate manufacture and importation volume, screening data on broad use categories, additional data on more detailed subuse categories, physical chemical properties, bioaccumulation, and persistence (biodegradation) (Fukushima 2012; Hirai 2012; Nanimoto 2012). If available, the government may also consider monitoring data, data from the Pollutant Release and Transfer Register, and information on handling. Authorities may request industry to submit necessary information if it is not already available. After conducting a risk assessment that utilizes this additional exposure information in combination with available hazard data, authorities attempt to make a risk management decision. If available hazard information is not sufficient for the ministries to reach a risk management decision, then they may require manufacturers and importers to submit hazard data on long-term toxicity and, if necessary, conduct tests to generate hazard data where none are currently available.† After risk assessment, authorities designate a PACS as a General Chemical Substance if the assessment shows that risk is sufficiently low, a Monitoring Chemical Substance if it meets the criteria for persistence and

---

* CSCL, Art. 9(1).
† CSCL, Art. 10.

bioaccumulation, or a Class II Specified Chemical Substance if the risk of harm to human health or the environment is reasonably likely.*

D. *Monitoring PB and Class I PBT substances*

"Monitoring Chemical Substances" are those for which persistence and bioaccumulation have been established but for which long-term toxicity remains uncertain.† In other words, Monitoring substances are PBs, an example of what we call "partial PBTs." Monitoring substances are considered candidates for the Class I designation as PBTs; however, the CSCL applies risk management to Monitoring Substances as well. As of October 2014, there are 37 Monitoring Substances (Takahashi 2014). Manufacturers and importers of Monitoring Substances must notify METI of the quantity of the listed substance that they produced or imported in the preceding fiscal year‡ and must also inform companies down the supply chain of the chemical's status as a Monitoring Substance.§ Additionally, the government may compel manufacturers and importers to conduct hazard assessments of the substance if it suspects that it may be harmful to human health or the environment.¶ The ministries may then use this information to determine whether or not to classify the chemical as a Class I Substance.**

The CSCL designates PBTs as Class I Specified Chemical Substances and prohibits their manufacture, importation (alone or in products), and use.†† As with its predecessors, the amended CSCL continues to define toxicity in terms of long-term or chronic toxicity, not acute toxicity. At present, there are 30 Class I substances.‡‡ Whereas prior legislation completely prohibited Class I PBTs, the amended CSCL allows companies to apply to METI for use-specific exemptions for "essential uses" (*Chemical Watch* 2011b; Uyesato et al. 2013).§§ The legislation specifies that METI may approve an application for manufacture or import only if the applicant shows that the quantity to be manufactured or imported is both necessary and no greater than that needed to meet domestic demand for a particular use.¶¶ Applicants must also demonstrate that their equipment and safety procedures meet specified standards and that they possess the fiscal and technical capabilities to adequately control risks. Finally, applicants must

* CSCL, Art. 11.
† CSCL, Art. 2(4).
‡ CSCL, Art. 13.
§ CSCL, Art. 16.
¶ CSCL, Art. 14.
** CSCL, Art. 15.
†† CSCL, Art. 2(2). For relevant regulations on Class I substances, see CSCL, Art. 17–34.
‡‡ Japan announced the addition of endosulfan and hexabromocyclododecane to its list of Class I substances on July 5, 2013 (*Chemical Watch* 2013).
§§ CSCL, Art. 17–26.
¶¶ CSCL, Art. 20 (manufacture) and 23 (importation).

show that substitute chemicals are difficult to obtain for the desired use and that the use is not likely to present a risk to human health or the environment.* Upon approval, Class I substances must be strictly handled and labeled, and together, the ministries may recall any chemical or product if they determine that it poses a risk.

Perfluorooctane sulfonate (PFOS) is the only Class I substance for which METI has approved use-specific exemptions (Ikemoto 2011). The exemptions allow for PFOS or its salts to be used in the manufacture of semiconductors and for the manufacture of professional-use photographic film.† No other exemptions have been granted, and many consider the Class I designation to be a de facto ban on the manufacture, importation, and use of the chemical (Toda 2007; Uyesato et al. 2013).

E.   Class II substances

Class II substances under the pre-2009 law were those that are persistent and toxic, but not bioaccumulative, another example of a partial PBT designation. Under the new legislation, however, Class II substances are those that are hazardous to humans, plants, or animals "due to a considerable amount of the chemical substance remaining in the environment over a substantially extensive area or because it is reasonably likely that such a situation will arise in the near future in view of its properties and manufacture, import, use, etc."‡ This category, therefore, still includes PT substances, but it also seems to include substances that are subject to continuous release, including chemicals that are neither persistent nor bioaccumulative, but for which exposure might be harmful because of toxic effects from, for example, CMR or endocrine disrupting properties (Naiki 2010; Kogan 2013; Uyesato et al. 2013). As of 2014, there are 23 Class II substances, which were all listed before this change. Thus, the scope of the amended Class II designation remains uncertain. Once designated as Class II, chemicals are subject to quantity notification requirements not only for the prior year but also for planned quantities for the forthcoming fiscal year, which act as caps for the manufacture and importation of the substance.§ The CSCL also compels the ministries to develop and apply technical guidelines for production, handling, and labeling of Class II chemicals.¶

---

* CSCL, Art. 25.
† Order for Enforcement of the Act on the Evaluation of Chemical Substances and Regulation of Their Manufacture, etc., Cabinet Order No. 202 of June 7, 1974, Art. 8.
‡ CSCL, Art. 2(3). For relevant regulations on Class II substances, see CSCL, Art. 35–37.
§ CSCL, Art. 35.
¶ CSCL, Art. 36–37.

F.   Restriction authority

The CSCL, like REACH, provides general restriction authority. Article 38 of the amended CSCL allows the ministries to recommend targeted restrictions on the manufacture, importation, and use of chemicals that are not designated Class I or Class II but that the ministries suspect satisfy the criteria for designation as Class I or Class II. As this risk management approach has not yet been widely used, it is too early to assess the efficacy of the practice.

G.   Risk management principles

The CSCL utilizes the PBT concept for each of the three policy types we have identified. First, the CSCL's assessment regime has traditionally been focused on *gathering information* on PBT characteristics, although recent amendments have broadened the scope of the CSCL to encompass additional factors: exposure-related information and additional toxicological endpoints such as adverse effects on the endocrine system. Second, the amended CSCL utilizes PBT characteristics, in addition to other factors, to *prioritize* chemicals for assessment and management. Third, the CSCL treats the PBT designation as a trigger for *risk management.* The way in which the CSCL utilizes the PBT concept, however, unevenly adheres to the risk management principles.

Whereas the initial CSCL was entirely hazard based, the amended CSCL was designed with a risk-based approach in mind. The use of the PBT designation in the CSCL retains its basis in hazard and might not adhere to the principle of differentiation of uses to the extent that the PBT Class I designation functions as a de facto ban. Moreover, the use of exposure criteria to screen chemicals before assessment should mean that any chemicals designated Class I in the near future would have already met those risk-based screening criteria.

In a positive advancement, the amended CSCL incorporates exposure-related information into the assessment and management processes. Importation and production volumes are used as screening criteria to determine which chemicals are subject to tiered risk assessment and, hence, potential risk management. The PACSs prioritization and tiered risk assessment process serve as mechanisms to inject information on use and exposure into the regulatory process. The PACSs risk assessment process is ultimately a tool to prioritize chemicals for risk assessment, similar to Canada's categorization and CMP, discussed later in this chapter. After risk assessment, the management options available to Japanese authorities under the CSCL are somewhat limited, as we explain in the following paragraphs.

To manage risks under the CSCL, the ministries must designate substances as Class I, Monitoring, or Class II Chemical Substances. Any chemical that the ministries formally determine fits all three PBT criteria will be designated a Class I Chemical Substance, subject to a de facto ban on use, production, and importation. However, the amended CSCL does allow for essential uses. The ministries may consider risks associated with particular uses, socioeconomic information, and the availability of alternatives in making decisions on whether to allow a particular use of a PBT. However, to date, the Class I designation is seen as a de facto ban in practice (e.g., Uyesato et al. 2013).

Of the 30 recognized PBTs, PFOS is the only chemical that has been approved for an essential use. In addition to the option of granting approved exemptions for particular uses, Article 38 allows the ministries to recommend targeted restrictions on chemicals that satisfy the PBT criteria. At present, it is too early to comment on how these amendments will affect CSCL's risk management processes in practice. To the extent that a de facto ban remains the only employed risk management route for PBTs in practice, the CSCL's utilization of the PBT concept is not fully consistent with the principles of differentiation of uses and exposure and risk assessment.

Moreover, the application of risk management obligations to PB substances, labeled Monitoring Chemical Substances, is similar to the European approach toward very persistent, very bioaccumulative (vPvB) chemicals. In both cases, authorities treat chemicals that exhibit persistence and bioaccumulation but for which toxicity is indeterminate as chemicals of high concern. The Japanese approach in the CSCL, however, does not treat PBs and PBTs as regulatory equivalents. The Monitoring PB substances are subject to *informational* risk management requirements (e.g., manufacturers must notify downstream users that a chemical has been designated a Monitoring Chemical Substance). This is similar to placement of a substance on the Candidate List under REACH and may have a similar market deselection effect. However, REACH allows for European authorities to treat vPvB and PBTs as regulatory equivalents—both may be considered substances of very high concern (SVHCs) and placed on the Candidate List and then the Authorization List. The CSCL does not treat PBs and PBTs the same, though, since a Class I designation requires a finding of long-term toxicity.

The difference might be a product of the flexibility of regulatory risk management options. Given that the Class I designation might operate as a de facto ban, it is desirable that a toxicity finding is necessary before the application of risk management. If the risk management approaches under the CSCL become more flexible in practice (e.g., through exempted uses for which the benefits outweigh the risks or through use-based restrictions), then the regulatory classification becomes less important. Under European, Canadian, and American systems, the regulatory label does not necessitate the application of any particular risk management option, so the implication of labeling a chemical a PBT or partial PBT has less formal regulatory impact. The fact that the CSCL provides authority for targeted risk management that may be applied differentially to particular high-risk, low-benefit uses is promising. Furthermore, the amended CSCL broadens the Class II category from PT chemicals to account for exposure and additional hazard endpoints. This is another positive advancement that reflects the broader principle that risk management decision making should not be driven solely by hazard-based labels.

Whereas the CSCL unevenly adheres to the differentiation and exposure and risk assessment principles, it does adhere to the priority setting, value of information, and precautionary principles. The amended CSCL includes a prioritization process (the PACS designation) that uses exposure and PBT characteristics (as well as others) to screen chemicals into the risk assessment and management framework and then further prioritize those chemicals for assessment and management decisions. The risk assessment processes are tiered, and a lack of information at a particular tier in the process does not constitute a barrier for decision making. For example, in the prioritization process, a lack of information on a particular toxicological endpoint generally results in an automatic application of a high hazard class. This is an example of precautionary decision making built into the design of the risk assessment process. In this way, full information and comprehensive risk assessments utilizing data from the field and experiments are not necessary for regulatory decision making. That decisions may be made based on limited data suggests that the CSCL is consistent with the value of information principle.

The extent to which the CSCL adheres to the information quality and rational alternatives principles is inconclusive as of yet. Overall, the Japanese system, as it is formally designed, seems strongly rooted in the precautionary and priority-setting principles. The CSCL makes only limited use of the exposure and risk assessment principle as applied to PBTs, given that the designation carries risk management implications. Exemptions are permitted to honor the differentiation principle, but socioeconomic analysis and alternatives analysis are not embraced as explicitly as they are under REACH. On the other hand, the process of implementation of the CSCL may be more flexible than the formal design of the system suggests. Japanese regulators are inclined to participate in international processes, whenever feasible. It is also too early to assess how much the Japanese reforms of 2009 have modified the reality of the CSCL regulatory system. Additional studies on the CSCL on its own and in comparison with other regulatory programs are warranted.

III. China

A. Chemical manufacturing and regulation in China

China is home to the world's largest chemical manufacturing industry as well as a substantial industry for consumer goods production. As one of the world's manufacturing centers and with a large and growing population, China faces difficult environmental and public health challenges. The challenges posed by the release of organochlorine pesticides and industrial chemicals, PBTs in particular, in China have been well documented (e.g., Wei, Kameya, and Urano 2007; Li et al. 2011). Although the production and use of POPs have been restricted for decades, communities around China continue to face a legacy of exposure to POP pesticides, including dichlorodiphenyltrichloroethane (DDT) and hexachlorocyclohexane (HCH) (Hu, Zhu, and Li 2007; Wei, Kameya, and Urano 2007) as well as industrial chemicals and their by-products such as dioxins, furans, PCBs, and polybrominated diphenyl ethers (PBDEs) (Zhang et al. 2007; Chen et al. 2012; Ni et al. 2013). Recently, the documentation of cancer clusters in areas with high chemical pollution, dubbed "cancer villages" in the popular media, has brought attention to the exposure of communities to chemicals, including PBTs (Li et al. 2011; McKenzie 2013; Tremblay 2013).

At present, there are over 45,000 chemicals in commerce listed on the Inventory of Existing Chemical Substances in China (Liu et al. 2012; Kexiong 2014). Many industrial chemicals and pesticides began to be used in the 1950s, as they were elsewhere in the world. However, China's regulatory framework to govern cradle-to-grave management of these chemicals did not develop alongside those in Japan, Canada, the United States, European nations, and other industrialized countries. Only recently has China begun to develop a regulatory system for chemicals.

China's initial concerns centered on POPs (especially pesticides). In 1979, the Standing Committee of the People's Congress enacted the Environmental Protection Act of the People's Republic of China, which encouraged use of lower-toxicity pesticides over higher-toxicity pesticides (Wei, Kameya, and Urano 2007; Lau et al. 2012). The Chinese government restricted the production, importation, and use of DDT and many of the other dirty dozen POPs in the early 1980s, but it gave relatively little attention to industrial chemicals. When industrial chemicals drew attention, regulators tended to focus on individual chemicals rather than classes of chemicals. For example, in 1991, China enacted a regulation that restricts the importation, production, and use of PCBs in electrical equipment and governs PCB disposal (Lau et al. 2012).

There are now a myriad of laws that are relevant to industrial chemicals and various PBTs (Zhang et al. 2005; Wei, Kameya, and Urano 2007; Wang et al. 2012; Uyesato et al. 2013). The literature on Chinese chemical regulation recognizes some difficulties in studying these laws. First, the Chinese legal framework creates difficulties to those unfamiliar with the Chinese governance structure. Chemical laws have been enacted at multiple levels of government and include acts (formal products of legislative bodies), regulations (authoritative rules or orders issued by the State Council), provisions (rules or orders issued by competent authorities), and standards (more specific implementation rules issued by competent authorities) (see Zhang et al. 2005; Wei, Kameya, and Urano 2007). Second, authority to implement chemical regulations is distributed among nearly a dozen different governance bodies, including the Ministry of Environmental Protection (MEP, formerly the State Environmental Protection Administration), the State Administration of Work Safety (SAWS), the General Administration of Quality Supervision, the Inspection and Quarantine, the National Health and Family Planning Commission, the Ministry of Industry and Information Technology, the Ministry of Public Safety, the National Development and Reform Commission, the China Food and Drug Administration, the Ministry of Transport, the Ministry of Agriculture, the State Administration for Industry and Commerce, and the General Administration of Customs. No formal mechanism exists to coordinate action by these administrative bodies. In fact, the lists of chemicals targeted by different agencies under different laws often overlap with little organized synchronization (Wang et al. 2012; Uyesato et al. 2013). However, certain laws do delegate authority to particular agencies. Several publications on Chinese governance of POPs have created lists that make sense of the tapestry of laws that apply to various chemicals and classes of chemicals (e.g., Zhang et al. 2005; Wei, Kameya, and Urano 2007; Lau et al. 2012; Wang et al. 2012).

Despite the number of laws that Chinese authorities have enacted, the impact of these laws is uncertain because of endemic enforcement and implementation problems, inadequate technical resources for risk assessment, and the lack of guidance for stakeholders and public authorities (Zhang et al. 2005; *Chemical Watch* 2011b; Lau et al. 2012; Wang et al. 2012). More recently, there has been a wave of new legal advances, which have together created momentum for the development of a comprehensive risk assessment and management framework in China (Liu et al. 2012).

In particular, the Chinese government updated its new chemical regulation in 2010, when it enacted the Regulations on Environmental Management of New Chemical Substances. This regulation is often referred to as "China REACH" (Uyesato et al. 2013). The law establishes a risk-based registration program for new chemicals (Lau et al. 2012; Wang et al. 2012), but it does not establish an entirely REACH-like system. It is not as comprehensive as REACH in the scope of chemicals covered nor does it provide authorities with the risk management tools that are available to European authorities under REACH. Nonetheless, Chinese risk assessment guidelines and regulatory infrastructure have emerged and are likely to facilitate the implementation of regulations on existing chemicals as well (Lau et al. 2012; Liu et al. 2012; Wang et al. 2012).

Here, we discuss the use of the PBT concept in some of China's primary laws on existing chemicals: the Environmental Management Provisions on the First Import of Chemicals and the Import and Export of Toxic Chemicals (Toxic Chemicals Regulation), the Regulations on the Safe Management of Hazardous Chemicals (Hazardous Chemicals Regulation),* and the Twelfth Five-Year Plan for Chemical Environmental Risk Prevention and Control.

B. Import and export of "toxic" chemicals

The first way in which China utilizes the PBT concept is through its registration regulations, which serve as information gathering and risk management systems. One of China's first modern laws focusing on industrial chemicals was the Environmental Management Provisions on the First Import of Chemicals and the Import and Export of Toxic Chemicals, issued by the National Environmental Protection Agency (now the MEP), the General Administration of Customs, and the Ministry of Foreign Trade and Economic Cooperation. Issued in 1994, the law fulfills China's obligations under the London Guidelines for the Exchange of Information on Chemicals in International Trade (Lin 2012; Wang et al. 2012; Uyesato et al. 2013). Amended in 2007, the law established a registration system for the importation and exportation of designated toxic chemicals. Before importing or exporting designated chemicals, firms must obtain a certificate from the MEP. To register, companies are required to pay a fee, submit information on a chemical's hazards to the MEP, and provide the MEP with a sample of the chemical for testing.

The registration obligations apply to chemicals that are on the List of Toxic Chemicals Severely Restricted for Import or Export by China (Wang et al. 2012; Uyesato et al. 2013). The initial versions of the list included 27 chemicals, including many of the dirty dozen POPs. The most recent version of the list (2014) includes 162 chemicals.† In addition, authorities have issued several lists that ban the import and export of certain chemicals,

---

* State Council Decree No. 591 (passed on February 16, 2011, published on March 2, 2011, and entered into force on December 1, 2011) (hereinafter Decree No. 591). The Chemical Inspection and Regulation Service provides a translated English version of Decree No. 591 at http://www.cirs-reach.com/China_Chemical_Regulation/Regulations_on_safe _management_on_hazardous_chemicals_China_2011.pdf.

† Early versions of this list are available online in English at http://english.mep.gov.cn /inventory/. The Chemical Inspection and Regulation Service has made the 2014 edition of the list available at http://www.cirs-reach.com/China_Chemical_Regulation /Registration_of_import_export_of_toxic_chemicals_in_China.html.

primarily POPs listed under the Stockholm Convention (Hu, Zhu, and Li 2007).

The law defines "toxic chemicals" as those that can harm human health and the environment "through environmental accumulation, bio-accumulation, biotransformation, or chemical reaction, or cause serious hazards with potential physical risk through contact with the human body" (Uyesato et al. 2013, p. 369).* The inclusion of environmental accumulation, bioaccumulation, and harm through exposure suggests that the Chinese authorities had the PBT characteristics in mind as they defined the scope of the import and export restrictions. On the other hand, many of the listed chemicals are not PBTs or even partial PBTs, so the definition is sufficiently broad to allow authorities a wide scope of discretion in making listing decisions. Recently, the import and export registration system has been incorporated into China's broader registration system for existing chemicals.

C. Management of "hazardous" chemicals

In 2011, the State Council issued Decree No. 591, the Regulations on Safe Management of Hazardous Chemicals. Implementation authority is the primary responsibility of the SAWS.† The decree defines hazardous chemicals as "highly toxic substances and other chemicals which are toxic, corrosive, explosive, flammable or are combustion-supporting and can do harm to people, facilities or the environment."‡ A 2013 draft version of the Catalogue of Hazardous Chemicals included about 3000 chemicals and chemical groups (Liu 2014). Like chemical regulations in other jurisdictions, the definition of chemicals within the scope of Decree No. 591 is broader than those with PBT characteristics.

Decree No. 591 requires that companies that produce, use, operate (sell and distribute), and transport listed hazardous chemicals register listed chemicals and obtain licenses from local Chinese authorities.§ Producers must generate and provide Material Safety Data Sheets and are also responsible for precautionary labels in accordance with the GHS.¶ Firms are prohibited from purchasing listed chemicals from unlicensed operators or without safety data sheets and proper labels.** Decree No. 591 also includes provisions governing the handling, transportation, and storage of listed hazardous chemicals as well as provisions governing the

* Environmental Management Provisions on the First Import of Chemicals and the Import and Export of Toxic Chemicals, Art. 4(4).
† Decree No. 591, Art. 6(1).
‡ Decree No. 591, Art. 3.
§ Decree No. 591, Art. 14 (production), 29 (use), 33 (operation), 43 (transportation).
¶ Decree No. 591, Art. 15.
** Decree No. 591, 591, Art. 37.

preparation of plans to respond to accidental releases.* The decree also mandates that manufacturers and importers of listed chemicals register with the National Chemicals Registration Centre of the SAWS. Registrants must submit information on classification and labeling, physical and chemical properties, hazardous properties, uses, safety measures for storage, use, and transportation, and emergency response measures in the case of accidents.[†]

In October 2012, under the authority of Decree No. 591, the MEP published Order No. 22, Measures for the Registration of Hazardous Chemicals for Environmental Management. The Order entered into force in March 2013. In addition to the general requirements of Decree No. 591 and supporting measures issued by the SAWS, MEP Order No. 22 imparts more rigorous registration and management obligations on specified Hazardous Chemicals for Priority Environmental Management (priority chemicals).[‡] Producers and users must register with the designated authority and receive a certificate for production and use.[§] Registration must include information on the registrant company, environmentally sensitive areas in the region of use or production, the quantity of the chemical to be used or produced, hazardous categories, uses, the safety data sheet, risk management measures, releases, wasted disposal methods, an environmental impact assessment, emergency response plans, and an environmental monitoring report (Kexiong 2014).[¶] Moreover, producers and users must contract with an approved third party to conduct a risk assessment and submit the risk assessment with the registration.[**] A registration certificate is valid for a period of three years and must be renewed thereafter.[††] Registered producers and users must submit a yearly "release and transfer" report, including information on releases, disposal, and recycling, and also a pollution prevention and control management plan, including information on measures to reduce releases of the chemical, treatment plans, and risk-reduction capacity building.[‡‡] Companies must continually monitor releases,[§§] provide designated information on releases to the public,[¶¶] maintain records,[***] and provide for inspection,

---

[*]   For an overview of the requirements of Decree No. 591, see Liu (2014) and Thompson et al. (2011).
[†]   Decree No. 591, Art. 67.
[‡]   MEP Order No. 22, Art. 3.
[§]   MEP Order No. 22, Art. 6–8.
[¶]   MEP Order No. 22, Art. 9.
[**] MEP Order No. 22, Art. 10–11.
[††] MEP Order No. 22, Art. 13–14.
[‡‡] MEP Order No. 22, Art. 20.
[§§] MEP Order No. 22, Art. 21.
[¶¶] MEP Order No. 22, Art. 22.
[***] MEP Order No. 22, Art. 23.

monitoring, and government supervision.* Additionally, MEP Order No. 22 incorporates the import and export restrictions regime into the Decree No. 591 and Order No. 22 framework.†

Priority chemicals constitute only a relatively small subset of the listed hazardous chemicals under Decree No. 591. MEP Order No. 22 does not specify the criteria by which priority chemicals are to be identified, but the MEP released a draft list of 142 chemicals in April 2013, indicating that selection criteria include PBT and vPvB characteristics, acute or chronic toxicity to the aquatic environment, serious hazards to human health (e.g., CMR chemicals), chemicals subject to international environmental conventions, chemicals that are prone to accidental releases, and any other chemicals for which risks need to be reduced (CIRS 2014). Some sources indicated that PBT and vPvB criteria might play a central role in identifying priority chemicals (e.g., Wu 2012; Du 2014; Zhou 2014), but this has not occurred so far. In 2014, the MEP released a final version of the first list of priority chemicals, including 84 chemicals and relatively few PBT/vPvB chemicals (Du 2014; Kexiong 2014). Accounts differ as to the role that PBT characteristics might play in the future of priority chemicals identification under MEP Order No. 22 (Chynoweth 2014; Du 2014).

D.   Risk management under the Twelfth Five-Year Plan

Whereas Decree No. 591 and MEP Order No. 22 impart registration and risk assessment and management obligations on industry, they do not provide the government with the authority to issue restrictions on chemicals or uses. Thus, alongside the registration and management regimes under Decree No. 591, the MEP in February 2013 announced an additional regulatory effort to manage toxic chemicals: the Twelfth Five-Year Plan for Chemical Environmental Risk Prevention and Control (CIRS 2013; Shengxian 2013).‡

Under the Twelfth Five-Year Plan, the MEP announced that it intends to establish two lists by the end of 2015: an elimination list and a restriction list. The plan identifies 58 chemicals and chemical types (some overlapping with the list of Hazardous Chemicals for Priority Environmental Management) and places them into three groups: 25 that present "accumulative risks," including cyclohexane, dichloroethane, bisphenol A, trichlorobenzene, and hexachlorobutadiene (HCBD); 15 that are the subject of frequent accidents, including benzene and aromatics, cyanide, chlorine, and methanol; and 30 "characteristic pollutants" (12 of which are listed

---

* MEP Order No. 22, Art. 24–25.
† MEP Order No. 22, Art. 16–19.
‡ Guo Fa [2011] No. 42. An unofficial English translation is available from Greenpeace at http://www.chemsec.org/images/stories/2013/news/Priority_Chemical_List_for_Prevention_and_Control_Outlined_in_Chinas_12th_FYP.pdf.

in the frequent accidents category as well), including formaldehyde and benzene (ChemSec 2013; CIRS 2013; Lin 2013). The use, manufacture, and import of most of these chemicals are already regulated or prohibited in other industrialized nations.

By the end of 2015, the MEP plans to be monitoring all of these chemicals. The MEP reported that it selected these 58 substances for monitoring and potential risk management because of their PBT, vPvB, CMR, and endocrine-disrupting properties (CIRS 2013; Liu 2013; Tremblay 2013). The plan will also target seven industry sectors for priority management: chemical raw materials, pharmaceuticals, petroleum refining, coking and nuclear fuel processing, smelting and pressing of nonferrous metals and textiles, chemical fibers, and coal-based chemicals (CIRS 2013). No restrictions have been enacted as of 2014.

E.   Risk management principles

Only limited information is available about the details of China's new risk assessment and managements systems. Decree No. 591, MEP Order No. 22, and the Twelfth Five-Year Plan are in their early years of implementation. It is therefore not feasible to fully judge the Chinese system based on our risk management principles.

Several risk management obligations are unclear, and implementation details are continuing to develop (Chynoweth 2014). For example, it is unclear whether Chinese authorities may reject a registration: Some suggest that the registration system under MEP Order No. 22 is simply intended to provide the government with information that can later be used to issue further risk management if necessary (Arputharajoo-Middleton 2014). On the other hand, given the nature of the particular requirements under MEP Order No. 22, it seems that registering a priority chemical for production or use might carry some significant risk management obligations. The design of MEP Order No. 22 is somewhat similar to the REACH registration system in that it shifts the burdens of production of information and proof of safety on to industry. It is uncertain at present how the regulations will unfold in practice.

Moreover, it is also unclear how Chinese authorities define PBTs and vPvBs, and how the MEP used these determinations to establish the lists of 84 priority chemicals under Order No. 22 and 58 types of chemicals under the Twelfth Five-Year Plan. Nor is it clear what weight was given to the PBT criteria in the MEP's decision making relative to other considerations. And it is not clear whether there is any role for differentiation of specific uses or alternatives assessment in the Chinese system. The definitions of toxic chemicals under the import and export restrictions, "hazardous" chemicals under Decree No. 591, and "priority" chemicals under MEP Order No. 22 seem to be sufficiently broad as to allow authorities a wide range of discretion in choosing which chemicals to list. Various accounts note that there does not seem to be any clear or consistent criteria that form the basis for listing decisions (e.g., Lin 2013; Arputharajoo-Middleton 2014; Chynoweth 2014; Du 2014; Zhou 2014). The formal PBT category does not seem to play a particularly substantial role in Chinese regulatory systems, or at least not as central a role as it plays in European and Japanese regulations. Nonetheless, the PBT concept does play some role in the way that Chinese authorities prioritize chemicals for risk assessment and management. Whether the Chinese system adheres to the risk management principles is worth additional investigation as the system is refined and made more transparent.

IV. Canada

A. *Canadian Environmental Protection Act of 1999*

Canada regulates industrial chemicals primarily under the authority of the CEPA, which was first enacted in 1988 and revised in 1999.[*] Like REACH and Japan's CSCL, Canada uses the PBT determination for both prioritization and risk management.[†] Specifically, PBT information, in addition to a finding of (at least potential) harmful exposure,[‡] is an input into the initiation of the risk management process. Unlike the European and Japanese programs, though, a positive PBT determination in Canada does not by itself trigger any immediate or certain risk management actions. Moreover, once Environment Canada (EC) or Health Canada (HC) determines that risk management is necessary, CEPA provides them the flexibility to implement a wide variety of risk management tools.[§] Under the framework established by CEPA, Canadian authorities have categorized chemicals in commerce to identify priorities for assessment, engaged in additional prioritization for identified chemicals of concern, conducted screening level risk assessments (SLRAs) to determine if risk management is warranted, and then made risk management decisions. This ongoing process is depicted in Figure 5.2.

---

[*] We use the acronym CEPA to identify the present law, enacted in 1999. Where we reference the former law, we identify it as CEPA 1988.
[†] For a comparison between the European REACH Regulation and the Canadian Environmental Protection Act of 1999, see Abelkop and Graham (2015), from which we adapt some background description, figures, and tables. See also Applegate (2008) and Denison (2007).
[‡] CEPA, § 64 (defining "toxicity" under CEPA).
[§] CEPA, § 93.

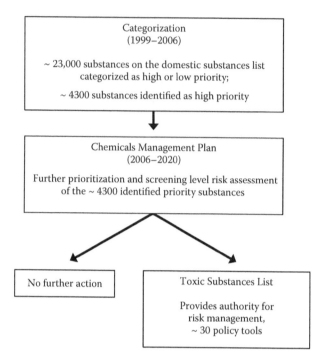

Figure 5.2 Overview of prioritization and risk assessment under CEPA. (From Abelkop, A.D.K., and J.D. Graham, "Regulation of Chemical Risks: Lessons for Reform of the Toxic Substances Control Act from Canada and the European Union." *Pace Environmental Law Review* 32, 2015.)

B. *Domestic Substances List categorization*

CEPA forms the basis for present regulatory activities by requiring a categorization and priority-setting process. The legislation required, by 2006, categorization of the roughly 23,000 chemicals on its Domestic Substances List (DSL) based on risk profiles—rudimentary analyses based on existing data. The initial DSL categorization identified roughly 4300 priority chemicals that warranted further assessment by government and industry. PBT properties played a significant role in this effort.

CEPA § 73(1) establishes persistence, bioaccumulation, and "inherent toxicity" (iT) to flora and fauna as its ecological criteria and "greatest potential for exposure" and iT to humans as its human health criteria for priority categorization.* In addition to searching for chemicals that meet all three PBiT criteria, Canadian authorities also searched for partial PBTs—chemicals that are iT and either persistent or

---

* CEPA, § 73(1).

bioaccumulative. The government of Canada enacted the Persistence and Bioaccumulation Regulations in 2000 to specify threshold criteria for identifying PBiT chemicals.* The thresholds are listed in Chapter 2 in Table 2.1.

Under the categorization, EC was responsible for assessing the ecological prioritization criteria, while HC was responsible for assessing human health. The iT determination, which denotes that exposure to a chemical may cause toxic effects, is equivalent to a toxicity determination in other contexts. It is solely a hazard-based determination of whether a substance causes toxic effects at tested doses. Canada uses the iT designation, though, because "toxic"—without the preceding *i* for "inherent"—has a specific legal meaning under CEPA that does not correspond to the general scientific understanding of toxicity (Meek and Armstrong 2007).

As an initial prioritization effort, Canada's categorization relied primarily on existing data. The DSL included about 23,000 chemicals that were manufactured in or imported into Canada in quantities equal to or greater than 100 kg/year between January 1, 1984, and December 31, 1986. EC and HC completed the categorization in September 2006, identifying 3856 chemicals that met the criteria for human health, ecology, or both (EC 2013a). Specifically, 197 chemicals met both the health and ecological criteria; 590 chemicals met only the health criteria; 3054 chemicals met only the ecological criteria; and 15 chemicals met the health criteria and had uncertain risk profile results for the ecological criteria.[†] Of the chemicals that met the ecological criteria, EC reports that the effort yielded a total of 393 potential PBiTs, 2047 potential PiTs, and 811 potential BiTs (EC 2013a).[‡] In addition, HC determined that another 345 chemicals, which did not meet either the human health or ecology criteria, nonetheless warranted further attention from a human health perspective (Easton 2008; EC 2013a). The results of the DSL categorization yielded about 4300 chemicals that warranted further assessment.

While there are similarities in how Canada, Europe, and Japan categorize substances, the categorization schemes are not identical. Table 5.2 displays how the Canadian, Japanese, and European regulatory programs

---

* Persistence and Bioaccumulation Regulations (Canadian Environmental Protection Act, 1999), SOR/2000-107.
† The categorization yielded only potential PBiT determinations because the risk profiles used for categorization did not include the same level of detail as the screening level risk assessments that EC uses to reach a decision on whether substances are PBiTs.
‡ Accounts of these figures vary depending on the way that different sources interpret the categories and the time at which they assessed the results (as certain chemicals may have shifted categories over time). For example, EC reported in 2008 that the categorization yielded 397 potential PBiTs.

Table 5.2 Various regulatory groupings of persistence, bioaccumulation, and toxicity

	Jurisdiction	Law	Designation
PBT	EU	REACH	SVHC (PBT)
	Japan	CSCL	Class I substance
	Canada	CEPA 99	PBiT
PB	EU	REACH	SVHC (vPvB)
	Japan	CSCL	Monitoring substance
PT	Japan	CSCL	Former Class II substance
	Canada	CEPA 99	PiT
BT	Canada	CEPA 99	BiT

employ different groupings of the persistence, bioaccumulation, and toxicity criteria to identify PBTs and partial PBTs.*

Following the initial DSL categorization, EC and HC examined industry data gathered from 2001 to 2006 to determine whether certain priority chemicals were still in commerce within Canada at or above the 100 kg/year DSL threshold. Of the potential PBiTs, 145 were identified that did not meet the criteria (EC 2008). EC removed these substances from priority consideration, but they are subject to requirements under the Significant New Activity (SNAc) approach, which governs the reintroduction and new uses of existing substances (Government of Canada 2012).† Therefore, about 250 potential PBiT chemicals remained for the government to examine through risk assessments. This value falls within the 100 to 1000 range of PBTs in commerce identified in Chapter 3. Accounting for the removal of 145 potential PBiT chemicals from priority consideration, the categorization still yielded more than 4000 chemicals for Canadian authorities to assess.

C.   *Risk assessment under CEPA*

Under CEPA, the determination of a chemical as a potential PBiT, PiT, or BiT in a categorization-level risk profile triggers an SLRA—an analysis that uses a weight of evidence approach to determine whether the chemical is toxic or capable of becoming so (Meek and Armstrong 2007; Easton 2008).‡ Recall that the determination that a chemical is toxic is distinct from whether the substance is "inherently toxic." The inherent toxic

---

* The United States is not included in Table 5.2 because the US EPA uses a scoring system that does not require formal PBT determinations for existing chemicals. China is excluded because it too does not make formal PBT determinations.
† CEPA, § 80.
‡ CEPA, § 74.

designation, in accordance with general understanding, is solely a hazard-based determination of whether a substance has toxic effects. *Toxic* under CEPA, on the other hand, is a purely legal term, which has been dubbed "CEPA-toxic." A substance is toxic under CEPA § 64 "if it is entering or may enter the environment in a quantity or concentration under conditions that" may result in harm to human health or the environment.* The CEPA-toxic designation, therefore, is more of a risk-based finding and is not solely based on hazard because it incorporates the potential for exposure. The determination that a chemical is CEPA-toxic authorizes the initiation of the risk management process.

CEPA and its predecessors also lay out a prioritization framework for risk assessment of industrial chemicals. As in the original CEPA 1988, the present legislation provides for the establishment of a Priority Substance List (PSL).† Under the PSL framework, EC and HC subjected listed substances to a more detailed risk assessment than an SLRA. Recall that SLRAs are only screening assessments. They may rely heavily on modeling and estimation techniques and conservative (high) estimates of exposure. Full risk assessments, on the other hand, may require the generation of data to determine, for example, modes of action and more likely exposure scenarios. However, both are designed to assess whether or not a substance is CEPA-toxic.

The first PSL, published in 1989, listed 44 chemicals (EC 2013b). Risk assessments were completed in early 1994 and identified 25 substances as CEPA-toxic (EC 2004). The government published the second PSL in 1995, this time listing 25 substances for risk assessment. Authorities conducted risk assessments on 23 of the chemicals and found 18 of them to satisfy the criteria for CEPA-toxicity (EC 2013c).

Canada has addressed a number of PBTs and other chemicals through the CEPA PSL framework, including dioxins, furans, hexachlorobenzene (HCB), HCBD, and chlorinated paraffins, to name a few (EC 2013b, 2013c). The government, industry, and NGOs considered the PSL process to be too slow and, ultimately, unworkable (Meek and Armstrong 2007). The length of the assessment process was a major driver for the creation of the 1999 update of CEPA, with its requirement for a seven-year period to categorize substances on the DSL. Notably, both PSLs were established under the original CEPA 1988 legislation. The current CEPA legislation maintains the PSL mechanism, but it has never been used under CEPA 1999, and it seems unlikely that it will be used in the near future. The PSL provides an informative contrast to the current system, the CMP.

* CEPA, § 64.
† CEPA, § 46(1)(a).

D.   Chemical Management Plan

As an alternative to the PSL, Canada introduced the CMP in 2006, after the completion of the DSL categorization (Harper 2006). The CMP is a strategy that is designed by EC and HC in cooperation with industry and nongovernmental organization (NGO) stakeholders. Its primary purpose is to assess and sustainably reduce the potential risks of chemicals in commerce, in accordance with the WSSD 2020 goals. A secondary purpose is to increase public confidence in industry's and the government's management of chemicals.

CEPA provides the primary legal authority for actions under the CMP. The CMP is designed to facilitate coordination between CEPA and other laws, including those that govern food and drugs, cosmetics, and pesticides. To that end, EC and HC draw legal authority for CMP actions from a variety of laws in addition to CEPA. Although many decisions have been contentious (see e.g., CSM and CELA 2009; Scott 2009, 2012), thus far, government, industry, and NGO stakeholders seem to be pleased with the design and progression of the CMP (Environmental Defence 2011). As such, the CMP has all but displaced the PSL as a prioritization mechanism for the assessment of chemicals in Canada.

Authorities are scheduled to work through the CMP in phases from 2006 to 2020. The phases are somewhat overlapping but also address some distinct sectors.

Phase I of the CMP included three primary programs. The first initiative of CMP Phase I was the industry "Challenge." The Challenge targeted nearly 200 of the chemicals identified in the categorization as high priority, including many potential PBiTs (e.g., PBDEs; see EC 2013d). EC and HC first divided the challenge substances into 12 "batches" to be addressed sequentially (Government of Canada 2011). CEPA § 71 provides government the authority to compel businesses to provide information about the chemicals that they manufacture, import, and use.* After releasing each batch of chemicals, EC and HC used this authority to direct industry to provide data on the chemicals in the batch. Much of this information consists of release and exposure data (Scott 2009). In some cases, however, additional hazard data were supplied. After receiving the data, EC and HC conducted SLRAs, which they released for public comment.

The ministries used the SLRA for each substance to determine whether or not it satisfied the criteria for CEPA-toxicity. When the assessment led the ministries to conclude that the substance is CEPA-toxic, they developed a risk management proposal, which they finalized after considering public comments. In addition to being a vehicle to determine whether risk management is necessary, the Challenge also encouraged

---

* CEPA, § 71.

companies to voluntarily reduce emissions of PBiTs and substitute, if possible, safer alternatives.

The second program of Phase I of the CMP was a Rapid Screening Assessment of potential PiTs and BiTs that are manufactured or imported in quantities no greater than 1000 kg/year (under the 1986 DSL), a total of 1066 substances (EC and HC 2013). EC evaluated whether these chemicals were already being assessed through other programs, searched for red flags by determining if the substances appear on priority or regulatory lists in other jurisdictions, and applied conservative ecological exposure scenarios to determine if further assessment is warranted. When the ecological exposure estimates were not of concern, HC applied a rapid screening framework from a human health perspective. Through this process, EC and HC determined that 472 potential PiTs and BiTs required further assessment and 533 required no further action because their estimated exposures were not of concern (EC and HC 2013).

The third program of CMP Phase I, which extends into Phase II, is the Petroleum Sector Stream Approach (Government of Canada 2014a). EC and HC divided 164 high-priority petroleum chemicals—some of which are potentially PBiTs—into five streams and have proceeded to gather information from industry, conduct SLRAs, and propose risk management options through the same processes as in the Challenge. As of 2014, Phase I is nearly complete, and authorities are working through Phase II.

Phase II was announced in 2011 and includes an additional rapid screening effort based on exposure-related information (EC and HC 2013),* an approach to address polymers, and the Substance Groupings Initiative (SGI) (Galatone 2013; Government of Canada 2014b). Under the SGI, EC and HC have placed an additional 500 substances into nine groups with similar chemicals—organic flame retardants, for example—and will proceed in the same manner as in the Challenge and the Petroleum Sector Stream Approach. The rationale for assessing substances in groups is that they can share similar chemical properties or are used in comparable ways. This approach emphasizes the use of the read-across technique, discussed in Chapter 2. Assessing like chemicals together therefore creates efficiencies for risk assessment and management and may facilitate the identification of safer substitutes. An overly aggressive use of this approach might stretch the limits of the read-across screening technique, which could undermine confidence in the government's SLRAs (e.g., CSM and CELA 2009).

Following the first two phases of the CMP, the Canadian government will still have to conduct SLRAs for about 1700 priority chemicals identified in the DSL categorization (Galatone 2013). How the ministries will

---

* To date, 117 substances have been identified that may not require further risk assessment because of low exposure potential (EC 2014).

execute the next phase of the CMP, which will begin in 2016, is uncertain; but it seems clear that, regardless of the outcomes of the next prioritization activity, the government will proceed in the same fashion as in the Challenge, Petroleum Sector Stream Approach, and SGI, with information gathering, screening assessment, and risk management.

E.   Risk management under CEPA

Each program under the CMP is designed to further prioritize chemicals to undergo SLRAs, which, as noted in Section C, are designed to determine whether or not a chemical satisfies the criteria for CEPA-toxicity. There are three potential outcomes if an SLRA leads authorities to determine that a chemical is CEPA-toxic.* First, the government may opt to take no further action. Second, the ministries may add the chemical to a PSL, triggering a more detailed and comprehensive risk assessment. This approach has been all but abandoned because full risk assessments have been deemed unnecessary for Canadian authorities to make a risk management decision. Third, the ministries may recommend that a chemical be added to Schedule 1 of CEPA, the Toxic Substances List (TSL) and, where applicable, the Virtual Elimination List (VEL) as well.

Not all outcomes of the SLRA process are discretionary. If the ministries determine that a chemical satisfies the PBiT criteria, that the release of the chemical "may have a long-term harmful effect on the environment," and that its presence in the environment "results primarily from human activity," then the ministries *must* recommend that the chemical be added to the TSL.† For any chemical recommended for inclusion on the TSL, whether discretionary or mandatory, the ministries are compelled to also recommend that the chemical be added to the VEL if it satisfies the criteria for persistence and bioaccumulation (but not necessarily iT), its presence in the environment "results primarily from human activity," and it is not naturally occurring.‡

1.   Toxic Substances List

The inclusion of a chemical on the TSL provides the ministries with the authority to propose and initiate risk management. As of December 2014, there are 132 chemicals or chemical types listed on the TSL.§ The addition of a substance to this list does not necessarily trigger the use of any specific risk management tool. Rather, CEPA provides EC and HC with a wide variety of risk management options to control exposure to

* CEPA, § 77(2).
† CEPA, § 77(3).
‡ CEPA, § 77(4).
§ The TSL is available online at http://www.ec.gc.ca/lcpe-cepa/default.asp?lang=En&n =0DA2924D-1&wsdoc=4ABEFFC8-5BEC-B57A-F4BF-11069545E434.

a hazardous chemical at any point in its lifecycle. Once an action is proposed, the ministries have two years to issue a regulation to adopt the proposed action.* CEPA § 93 provides authority for EC and HC to adopt any of about 30 different policy tools, including restrictions on the quantity of manufacture, sale, import, or export; amount, location, and conditions of releases; labeling, handling, and storage; and the generation and submission of information. The ministries may also issue guidelines, standards, or codes of practice or facilitate voluntary risk management efforts (Meek and Armstrong 2007).[†]

For example, authorities have issued regulations that pertain to specific classes of chemicals on the TSL, such as PBDEs and PCBs (EC 2013d, 2013e, 2013f); certain sources of TSL chemicals, such as pulp and paper mill effluent containing chlorinated dioxins and furans[‡] (EC 2013g); certain uses and products that contain TSL chemicals, such as concentration limits for 2-butoxyethanol in products for indoor use[§]; and more general risk management tools.

One such tool is the Prohibition of Certain Toxic Substances (PCTS) regulations.[¶] Authorities developed the PCTS regulations because "it was suggested that it would be simpler and more effective administratively to develop a generic banned-substances regulation to which substances would be scheduled rather than having separate regulations..." (EC 2013h). The PCTS regulations include several sublists, also called schedules. The 12 chemicals listed on Schedule 1 are prohibited from manufacture, import, sale, and use. Several POPs and PBTs are on this list, including DDT, Mirex, polybrominated biphenyl (PBB), HCB, and HCBD. Schedule 1 also prohibits an additional four chemicals unless they are manufactured within items.

PCTS Schedule 2 functions like REACH authorization and Japan's Class I determination: Listed chemicals are prohibited from manufacture, import, and sale, unless exemptions are provided under limited authority. Canada's exemption mechanism may be seen as more flexible than authorizations under REACH. The Minister of the Environment *must* issue a permit if "(a) there is no technically or economically feasible alternative... (b) the applicant has taken the necessary measures to minimize or eliminate any harmful effect of the toxic substance on the environment and human health," and (c) the applicant has prepared a plan to

---

* CEPA, § 91.
[†] CEPA, § 93.
[‡] Pulp and Paper Mill Effluent Chlorinated Dioxins and Furans Regulations (Canadian Environmental Protection Act, 1999), SOR/92-267.
[§] 2-Butoxyethanol Regulations (Canadian Environmental Protection Act, 1999), SOR/2006-347.
[¶] Prohibition of Certain Toxic Substances Regulations, 2012 (Canadian Environmental Protection Act, 1999) SOR/2012-285 (hereinafter PCTS Regulations).

phase out the use of the substance within three years after the permit is issued.* Schedule 2 lists five chemicals with permanent permitted uses, one chemical with a temporary permitted use, two with permitted concentration limits, and two with reporting thresholds. Thus, although the CEPA-toxicity standard does not necessarily mandate the consideration of socioeconomic data, consideration of alternatives, or differentiation in uses, such factors are built into the risk management decision-making process that follows a finding that a chemical is CEPA-toxic and the inclusion of the chemical on the TSL.

2.   Virtual Elimination List

If a chemical recommended for inclusion on the TSL is also determined to be persistent and bioaccumulative, anthropogenic, and is not a naturally occurring substance, then EC and HC must also recommend its inclusion on the VEL.[†] The concept of virtual elimination (as discussed in Chapter 4 in reference to the Great Lakes Water Quality Agreement) can be a source of confusion on whether detectable levels of a chemical in the environment must be virtually eliminated or whether detectable levels of a chemical in emissions must be virtually eliminated (Williams 2006). CEPA defines virtual elimination as the "reduction of a quantity or concentration...in the release below the level of quantification" prescribed by the ministries.[‡] That is, in the context of CEPA, virtual elimination refers to eliminating detectable levels of a chemical in emissions rather than in the environment. Once the level of quantification has been established, the chemical and the level must be added to the VEL. Authorities will then establish release limits for the chemical and issue additional risk management as judged necessary.[§] At present, there are only two chemicals on the VEL: HCBD and PFOS, both of which Canadian authorities have classified as PBiTs.[¶] Although virtual elimination is rarely invoked, hexabromocyclododecane might also be added to the VEL (and PCTS Schedule 1) in accordance with its risk management plan and CEPA timelines (EC and HC 2011).

CEPA provides EC and HC with a wide variety of options, but it does not always provide the most appropriate tools. Virtual elimination under CEPA refers to preventative reductions of chemicals in releases, but the Toxic Substances Management Policy of 1995 (which is still in place) explicitly targets anthropogenic PBiTs for virtual elimination from the

---

* PCTS Regulations, § 10.

[†] CEPA, § 77(4).

[‡] CEPA, § 65(1).

[§] CEPA, §65(3).

[¶] The VEL is available online at http://www.ec.gc.ca/lcpe-cepa/default.asp?lang=En&n =2F918AD5-1. The ministries did not specify a level of quantification for PFOS.

environment (EC 2013i, 2013j). EC has used this tool to issue remediation regulations for HCB, dioxins, and furans, for example (EC 2013k).

3.   *Alternative risk management options*

Several other risk management options are available to Canadian authorities to manage risks posed by the use of PBiTs and other chemicals. One risk management instrument that is gaining momentum is the SNAc requirement. A SNAc is very similar in concept to TSCA Significant New Use Rules (SNURs). SNAcs are applied to chemicals for which current uses are either extremely limited and well controlled, or for which quantities in current Canadian commerce are very low (Government of Canada 2012). The SNAc is applied to enforce notification of new or increased use (with an associated requirement to provide risk-related information as per a New Substance Notification), which allows regulators to conduct an updated risk assessment. The SNAc risk management tool has been applied to many potential PBiTs, including the 145 potential PBiTs that were removed from priority consideration in 2008 because data indicated that they were not in commerce at or above the 100 kg/year DSL threshold.

Under the CMP, authorities also search for the "best placed act" under which to initiate risk management. In other words, EC and HC may choose to pursue management under the Food and Drugs Act, the Pest Control Products Act, the Canada Consumer Product Safety Act, or the Fisheries Act in lieu of or in addition to CEPA (MacDonald and Castellarin 2013).

F.   *Risk management principles*

The government of Canada, through CEPA and the CMP, applies many of the same risk management tools to PBiTs as do the EU, Japan, the United States, and other nations. Canada also uses the PBiT concept for both priority setting and for risk management. But unlike the European and Japanese approaches, exposure is a crucial element in Canadian risk management decisions for PBiTs as well as other chemicals of concern. Risk management is triggered by a finding of CEPA-toxicity—a risk-based term of art that is defined by exposure as well as hazard. CEPA therefore adheres to the principle of exposure and risk assessment by institutionalizing the consideration of exposure data in its SLRAs and risk management decisions. That the ministries examine uses and exposure pathways as well as other applicable regulations (e.g., cosmetics and pesticide laws) in determining the appropriate risk management options, suggests that Canada's approach to chemical management adheres to the differentiation principle as well. CEPA does provide for the evaluation of alternatives, although formal comparative risk analysis is not institutionalized within the law.

The DSL categorization and CMP processes seem to adhere closely to the priority setting, exposure and risk assessment, and value of information principles. The CMP embodies the value of information and priority-setting principles by soliciting information on a specific, manageably sized group of chemicals of concern with strict deadlines for information submission. Rapid screening and the use of SNAcs embody the principles of priority setting, exposure and risk assessment, and value of information as well by focusing public resources for assessment on those chemicals that carry the most potential risk.

CEPA adheres to the value of information principle more strictly than REACH does because Canadian authorities do not compel industry to generate data unless the government has a potential concern about possible risk to human health, safety, and the environment. However, the REACH registration system is designed not only to facilitate industry-driven risk management; it is designed also to reduce the data gap on chemicals. Some criticize the Canadian approach for not fully reversing the burden of proof of safety on to industry (e.g., Briand 2010). The legislation does not require industry to make a safety determination, but CEPA does authorize EC and HC to compel industry to provide data in specific cases[*] and, in fact, this is an integral first step to the Challenge, Petroleum Sector Stream Approach, and the SGI.[†]

Moreover, the precautionary principle is a central guiding feature of the risk assessment and management decision-making process under CEPA. The legislation itself requires the application of both the precautionary principle and weight of evidence in decision making.[‡] The DSL categorization and SLRAs utilize a tiered risk assessment approach starting with upper-bounding exposure estimates and refining, as necessary and where possible, depending on the level of information available. A lack of information does not constitute a barrier to risk management decision making, as evidenced by the Siloxane D5 decision discussed in Chapter 3.

The spirit of the CMP is that it is a cooperative endeavor between government, industry, and NGO stakeholders. To be sure, praise of CEPA and the CMP is certainly not universal, as many specific decisions have raised controversy. Nonetheless, many stakeholders, including both industry and NGOs, seem to be pleased with the degree of activity under CEPA and the CMP, especially as compared with the level of activity before the enactment of CEPA 1999 (Environmental Defence 2011). As of 2014, Canadian stakeholders are not seeking to overhaul CEPA to the degree that stakeholders in the United States are seeking to overhaul TSCA (e.g., Hogue 2013). We turn now to a discussion of PBT policies in the United States.

---

[*] CEPA § 71.
[†] For additional comparison between CEPA, REACH, and TSCA, see Denison (2007), Applegate (2008), Renn and Elliott (2011), and Abelkop and Graham (2015).
[‡] CEPA § 76.1.

V.   United States

A.   Waste Minimization Prioritization Tool

In 1994, the US Environmental Protection Agency (US EPA) adopted the Waste Minimization National Plan (WMNP) under the authority of the Resources Conservation and Recovery Act (RCRA). RCRA establishes a cradle-to-grave management system for hazardous wastes. Under the WMNP, US EPA established a list of 31 Priority Chemicals and encouraged voluntary efforts to reduce the amounts of these chemicals in hazardous wastes (US EPA 2012a). Of the 31 chemicals, 27 are organic PBTs, including PCBs, HCB, HCBD, endosulfan, dioxins, and furans. In 2004, US EPA established the goal of reducing the release of PBTs in hazardous wastes by 15% of 2001 levels by 2008 (US EPA 2005). Other goals included avoiding the transfer of Priority Chemicals across media and reducing Priority Chemicals in discharges at their sources.

US EPA generated the Priority Chemicals list using the Waste Minimization Prioritization Tool (WMPT), which was designed to prioritize chemicals in commerce based on the risk they pose to human health and the environment (US EPA 1997). The WMPT uses a risk-based scoring process to determine an overall score for a chemical, which authorities can then compare to the scores of other chemicals to yield a relative ranking that prioritizes chemicals for risk assessment. The WMPT generates a score for each chemical, a derived value that has meaning only relative to the values for other chemicals. The overall score is based on assigned values for human health and ecological risk potential.

Risk potentials are determined by values assigned for exposure and toxicity. Exposure values for both human health and the environment are determined through scores for bioaccumulation potential, persistence, and production quantity. The human health toxicity value is based on scores assigned for cancer and noncancer effects. The ecological toxicity value is based on an aquatic toxicity score.

Values for each of the subfactors could be multiplied together to yield an overall score. After all, risk is often considered toxicity multiplied by exposure. Technically, though, persistence, bioaccumulation, and toxicity each come in different units, and their values vary across a wide range of degrees of magnitude, which means that a multiplication procedure yields a product that is hard to interpret. Explained in greater detail in Chapter 2, persistence is measured in half-life, bioaccumulation is typically measured with a bioconcentration factor (BCF), and toxicity is expressed as a parameter from a dose–response relationship. Regulators therefore encounter a practical problem when trying to determine the relative level of concern for different PBTs.

In a report for the Netherlands National Institute for Public Health and the Environment (RIVM), Rorije et al. (2011) draw a distinction among binning, ranking, and scoring prioritization systems. Binning systems distinguish between categories of chemicals to identify priorities. For example, systems that identify PBTs or CMRs as priorities relative to non-PBTs and non-CMRs utilize binning systems. A chemical is either in the PBT bin or not. Binning systems do not distinguish between chemicals within the designated categories. Ranking systems, on the other hand, generally use dense or ordinal ranking schemes to draw distinctions between individual chemicals. However, the distance between ranks is indeterminate, as there is no unit that can express the distance between a rank of 1 and a rank of 2 in the case of PBTs because persistence, bioaccumulation, and toxicity are expressed in different units. Nor can ranks be interpreted as absolute values. Rather, ranks only have meaning relative to one another. Finally, scoring systems for PBTs would generate separate scores for persistence, bioaccumulation, and toxicity that can then be compared with one another to prioritize chemicals. Scores for each of the characteristics can be generated on a continuous scale to allow for more meaningful comparison. Binning is the least resource-intensive approach, but it also yields the least amount of information. The inverse is true for scoring systems. The WMPT is an example of a scoring system.

The WMPT approach ranks chemicals corresponding to high, medium, and low scores for persistence, bioaccumulation, and toxicity (US EPA 1997). For example, a high bioaccumulation potential is associated with a BCF of 1000 or greater, the range from 250 to 999 is medium, and a BCF of less than 250 is low. The WMPT assigns scores based on available information or estimation techniques (if information is missing or inadequate). The tool assigns unit-less score values for each level: High BCF values receive a score of 3, medium BCF values receive a score of 2, and low BCF values receive a score of 1 (and so on for high, medium, and low persistence and toxicity). These proxy scores allow for the standardization of persistence, bioaccumulation, and toxicity into unit-less values.

Persistence, bioaccumulation, and toxicity, therefore, can each yield a score of 1, 2, or 3. These subscores are added to yield an overall score. Persistence and bioaccumulation are added to calculate the ecological and human health exposure potentials, which can range from 2 to 6. Scores for toxicity determine human and ecological toxicity, which can range from 1 to 3. Toxicity and exposure are added to calculate human health and ecological risk potentials, which can range from 3 to 9. Finally, ecological and human risk potentials are added to calculate the overall score, which can range from 6 to 18.

These ranges account only for persistence, bioaccumulation, and toxicity scores. The WMPT accounts for exposure data through the production quantity, which is calculated separately as a continuous variable and added into the equation as well. US EPA used the WMPT as a simple risk-screening tool to provide relative rankings as an initial step to risk assessment.

US EPA's approach to priority scores—one of the earliest attempted by a government agency—has many limitations: uneven data quality, arbitrariness in the relative weights assigned to different inputs, and little discrimination between chemicals as a result of broad categories. The importance of this initial effort, however, is not its technical validity or even its practical utility in waste management, but rather that it established a rudimentary model for scoring that was later refined by the US EPA under the TSCA, discussed in the following section. We are not aware of any effort to evaluate the consequences of the use of PBT criteria in the US EPA's 1994 waste management policy.

B.   *Emergency Planning and Community Right-to-Know Act*

The US Congress enacted the Emergency Planning and Community Right-to-Know Act of 1986 (EPCRA) to assist communities in emergency planning for releases of hazardous substances. EPCRA § 313 establishes the Toxic Release Inventory (TRI), which specifies reporting requirements for companies that dispose of toxic chemicals or release them into the environment. A key feature of the TRI inventories is that firms must disclose releases annually to the public in a format that is readily understood by reporters, community leaders, investors, and NGOs. One effect of TRI information, therefore, is to stigmatize facilities and firms with large volumes of emissions by providing information to consumers and investors (Hamilton 2005). Many facilities have therefore reduced rates of disposal and emission beyond what is required by federal and state regulation.

Historically, a facility was exempt from public reporting of TRI emissions if the extent of its emissions fell below a uniform value set by the US EPA. The exemption was widely used by small businesses to avoid reporting burdens.

In 1999, US EPA used PBT determinations as a regulatory justification to lower the information reporting thresholds under the TRI for certain PBTs.* For example, the release amounts that trigger a TRI notification requirement for PCBs, mercury, and lead were lowered from 10,000 lb (for users) and 25,000 lb (for manufacturers) to 10 lb, while the release threshold for dioxin and dioxin-like compounds became 0.1 g. Although many small businesses objected to this change, US EPA made the policy judgment that communities had a special right to know about emissions of PBTs, even at low levels. At present, the TRI list of PBTs includes 16 chemicals and four chemical compound categories (US EPA 2014a).

---

* Toxic Chemical Release Reporting: Community Right-to-Know, 40 C.F.R. § 372 (2013).

C. Toxic Substances Control Act

The TSCA of 1976 is administered by US EPA's Office of Pollution Prevention and Toxics (OPPT) and takes varying approaches to chemical regulation depending on whether a chemical is a new or existing chemical.* Although changes to TSCA have been proposed for many years, the prospects for legislative reform in the near future are uncertain.

1. New chemicals

Most of our analysis addresses regulations of existing chemicals. Yet, it is important to consider the regulation of new chemicals under TSCA because the legislation employs the PBT concept as part of its evaluation of new chemicals. TSCA § 5 requires firms to provide US EPA with a pre-manufacture notice before the production of a new chemical or the appropriation of an existing chemical to a new use, similar to the Canadian SNAc approach.†

US EPA reviews new chemicals and their planned uses to determine potential risks. This review process includes an explicit step for PBT assessment. The assessment takes a tiered approach where US EPA conducts an initial screening, a more in-depth screening on biodegradability and bioaccumulation, and finally, toxicity assessment and environmental fate testing. US EPA considers persistence, bioaccumulation, and toxicity criteria individually and in combination when making risk-based judgments.

There are two broad regulatory outcomes for new chemicals classified as PBT, depending upon the anticipated degree of risk. Under TSCA § 5(e), a determination of a moderate PBT classification requires exposure and release controls, followed by further testing to ensure risk is addressed. PBTs classified as "high risk concern" are denied commercialization unless further testing justifies removal from the "high risk" category.‡

Both the United States through TSCA and Canada through CEPA evaluate exposure data and exposure surrogates when considering the application of risk management to "significant new uses" of existing chemicals.§ TSCA § 5(a)(2) identifies factors that US EPA may consider in determining if a new use necessitates a SNUR, including projected manufacturing and processing volume, changes in the type or form of exposure, an increase in the magnitude or duration of exposure, and methods of manufacture, processing, distribution, and disposal.¶ Once US EPA issues

---

* For an overview of TSCA, see Auer and Alter (2007).
† TSCA, § 5; New Chemicals Program, 64 Fed. Reg. 60,194 (Nov. 4, 1999).
‡ Proposed Category for Persistent, Bioaccumulative, and Toxic Chemical Substances, 63 Fed. Reg. (Oct. 5, 1998).
§ TSCA, § 5(a); CEPA, § 80.
¶ TSCA, § 5(a)(2).

a chemical SNUR, firms must submit notice to US EPA 90 days before the manufacture, import, or processing of the chemical for that use. US EPA then has the opportunity to examine data on the new use before the firm actually begins using the substance in the new application. US EPA may compel risk management or disapprove the new use if necessary.

Through its SNUR authority, US EPA has addressed risks from PBT substances including PBBs,* several hundred perfluoroalkyl sulfonate (PFAS) derivatives,† PBDEs,‡ and various tetra and pentachlorobenzenes.§ While some of these actions predate recognition of the PBT concept, the bulk of existing SNURs are for PBTs.

2.   Existing chemicals

Three primary sections of TSCA govern existing chemicals. Under TSCA § 4, US EPA may compel chemical manufacturers, importers, and processors to generate and divulge new data on a chemical's adverse effects on human health and the environment. TSCA § 8 establishes requirements for companies to submit existing data on chemical identity, uses, production and importation volume, health and environmental effects, disposal, and potential exposure pathways. TSCA § 6 provides US EPA with the authority to apply various risk management instruments to chemical substances or uses that present an "unreasonable risk to human health or the environment" (see Auer and Alter 2007).¶

While TSCA utilizes a formal PBT determination for the management of new chemicals, the legislation does not instruct US EPA to make PBT determinations for existing chemicals. Rather, US EPA (2011) prioritizes existing chemicals for risk assessment by considering chemicals that are formally listed by other laws in the United States and elsewhere, and by generating separate scores for persistence, bioaccumulation, and toxicity as well as with other factors.

In August 2011, US EPA (2011, 2012b, 2014b) developed the TSCA Work Chemicals Plan, a two-step process for prioritizing existing chemicals for further risk assessment by OPPT. In the first step, US EPA identified candidate chemicals for review by searching various international, national, and subnational data sources for neurotoxins, PBTs, carcinogens, chemicals that might affect children's health or that are likely to be exposed to children, and chemicals detected through biomonitoring. US EPA relied on the TRI, the Great Lakes Binational Toxics Strategy, CEPA's DSL categorization, the Convention on Long-Range Transboundary Air Pollution, and the Stockholm Convention to generate its list of candidates with PBT

---

* Significant New Uses of Chemical Substances, 40 C.F.R. § 721.1790 (2013).
† Significant New Uses of Chemical Substances, 40 C.F.R. § 721.9582 (2013).
‡ Significant New Uses of Chemical Substances, 40 C.F.R. § 721.10000 (2013).
§ Significant New Uses of Chemical Substances, 40 C.F.R. §§ 721.1425, 1430, 1435 (2013).
¶ TSCA § 6(a).

properties. Across all six prioritization factors, US EPA compiled a list of 1235 candidate chemicals for assessment (2014b). From this initial group, US EPA removed from consideration those chemicals that are not subject to TSCA, already regulated, present radioactivity concerns, have complex process streams, are naturally occurring, and possess certain other properties. After removing these chemicals from the list, US EPA was left with 345 candidates.

In the second step to the prioritization process, US EPA assigned the remaining candidate chemicals three different scores: one for hazard (human and environmental toxicity), one for potential exposure, and a third combined score for persistence and bioaccumulation. The scoring system is a refinement of the system that was pioneered years earlier by the agency's waste policy program.

The scoring system is described in the OPPT's *TSCA Work Plan Chemicals: Methods Document* (2012b). Under the scoring system, each category (hazard, exposure, and persistence/bioaccumulation) is given a score of 1 to 3. The scores are then summed to yield an overall score, with high priorities ranging from 7 to 9; moderate priorities, 5 or 6; and low priorities, 3 or 4. If there is no hazard score or exposure score (because of missing data), but there is a PB score of 2 or 3, then the chemical becomes a candidate for additional information gathering.

The hazard score is determined from the highest score in any of the toxicity categories, which include acute mammalian toxicity, carcinogenicity, mutagenicity/genotoxicity, and so on. A chemical's hazard score can range from 1 (low) to 3 (high). To determine the hazard score, no independent testing is undertaken. Rather, data for particular endpoints are taken from existing sources.

The exposure score has three components, which each yields subscores: type of industrial use, general population and environmental exposure, and extent of release. The use type is based upon the amount of consumer use and likelihood for exposure. This score may range from 0 to 3, with 0 denoting no reported commercial use and a low likelihood of exposure. The second exposure criterion is the general population and environmental exposure score. A score of 3 is assigned for chemicals present in biota or measured in drinking water, indoor air, or house dust; a score of 2 for chemicals not detected in biota but reported present in two or more environmental media; and a score of 1 for chemicals detected in one environmental media. Finally, different release scores are given for TRI chemicals and non-TRI chemicals. TRI chemicals are scored based on the number of pounds released per year. Non-TRI chemicals are scored using production volume; the number of manufacturing, processing, and use sites; industrial processing and use; and commercial use. The subscores from each of these three exposure categories are then added and normalized to yield a final score of between 1 and 3.

Last, US EPA calculates a persistence/bioaccumulation score. As with the exposure categories, persistence and bioaccumulation are scored separately and then normalized (added together and then divided by 2). For persistence, a score of 3 is assigned to chemicals with a half-life greater than six months, a score of 2 for chemicals with a half-life from two to six months, and a score of 1 for chemicals with a half-life of less than two months. For bioaccumulation, a score of 3 is assigned to chemicals with a BCF/bioaccumulation factor (BAF) greater than 5000, a score of 2 for a BCF/BAF from 1000 to 5000, and a score of 1 for a BCF/BAF less than 1000. As with the hazard criteria, US EPA does not generate BCF and BAF values. Rather, regulators determine BCF and BAF values by analyzing other source material.

As noted above, the three component scores for hazard, exposure, and persistence/bioaccumulation are added to yield an overall score. Chemicals with a score of 7 to 9, denoting a high score, are given priority for risk assessment.

The scoring system is not without its shortcomings. For example, the scores do not provide an indication of the degree of uncertainty in the PBT determination process, as high uncertainty cannot be used to adjust an evaluation. Moreover, the ranking approach treats as equal PBT determinations made from estimation methods, field data, and laboratory tests. The scores have arbitrary scaling: Scores are unit-less values. Moreover, the equal weight given to the three component scores for hazard, exposure, and persistence/bioaccumulation is also arbitrary. One could plausibly argue that an unequal weighting process is at least of comparable validity.

Despite such concerns, US EPA is clearly striving to implement a system that adheres to the principle of priority setting. If errors in priority setting occur, the consequences are not irreversible. After all, prioritization does not determine which chemicals get assessed; it only determines which get assessed first. In accordance with the value of information principle, a prioritization system does not need to be perfect. It only needs to be good enough to provide regulators and stakeholders with a reasonable notion of which chemicals should be assessed first.

Using this approach, OPPT prioritized 83 chemicals for risk assessment (US EPA 2012b, 2014b). US EPA has since completed four assessments, removed over a dozen chemicals that are no longer in commerce, and added other chemicals to the Work Plan, including several flame retardants. OPPT will conduct full risk assessments on the three non-flame retardant chemicals (siloxane D4, 1-bromopropane, and 1,4 dioxane) as well as four of the flame retardant chemicals belonging to three groups (brominated phthalates, chlorinated phthalate esters, and cyclic aliphatic bromides) (US EPA 2013a, 2013b). OPPT will use data on the four flame retardants to inform its reviews of the remaining 16 flame retardants in each group, and it will consider data on 8 similar flame retardant chemicals outside the original 83 that it prioritized for review. As of October 2014, there are 90 chemicals or chemical groups on the TSCA Work Plan, including the following:

- 12 that received high persistence/bioaccumulation and toxicity scores,
- 26 that scored high for toxicity and moderate for persistence/ bioaccumulation,
- 5 that scored high for persistence/bioaccumulation and moderate for toxicity,
- and 2 that scored high for persistence/bioaccumulation and low for toxicity.*

US EPA's rationale for evaluating similar chemicals together is the same as that of EC and HC for the Petroleum Sector Stream Approach and SGI of the CMP: "Grouping and reviewing chemicals with similar characteristics together, rather than evaluating them individually, may help to address the potential for replacement of chemicals with well-characterized hazards by similar, but unevaluated, alternative substances with similar concerns" (US EPA 2013b). ECHA also intends to evaluate groups of substances (ECHA 2015).

Although US EPA uses persistence, bioaccumulation, and toxicity scores and the TRI list of PBTs to help set priorities for risk assessment under TSCA, the TSCA legislation includes no formal PBT policy for existing chemicals. TSCA's lack of a formal PBT construct for existing chemicals might be acceptable if the agency's regulatory program for existing chemicals were vibrant and rigorous like the Canadian program. However, US EPA, under TSCA, has made much less progress than EC and HC have in addressing concerns about chemicals with PBT and other properties of concern. A modernization of TSCA as a whole may be required before US EPA can be expected to effectively address existing chemicals of concern (whether PBTs or non-PBTs) (e.g., GAO 2013). Nonetheless, a strong feature of the TSCA and CEPA programs is serious treatment of the principle of exposure and risk assessment when determining whether chemicals with PBT properties should be regulatory priorities.

In addition, US EPA has been a leader in addressing concerns about PFOS and perfluorooctanoic acid (PFOA). For example, in the 2010/2015 PFOA Stewardship Program, companies committed to reduce global facility emissions and product content of PFOA and related higher homologue chemicals by 95% from 2000 levels by 2010 and to work toward eliminating

---

* The list of TSCA Work Plan chemicals and scoring results are available online (US EPA 2015).

emissions and product content by 2015 (US EPA 2013c, 2013d). The PFOA Stewardship Program is an example of how government can work with industry to reduce exposure to a chemical of concern.

D.   TSCA reform: Chemical Safety Improvement Act

In May 2013, the late Senator Frank Lautenberg (D-NJ) and Senator David Vitter (R-LA) released for public discussion a new bipartisan draft bill, titled the Chemical Safety Improvement Act (CSIA),* that would reform TSCA. For existing chemicals, the bill instructs US EPA to develop a new chemical assessment framework under which it would evaluate existing data, prioritize chemicals for safety assessments and determinations, and apply risk management to chemicals, which, under their intended conditions of use, do not meet the CSIA's safety standard. Safety assessments are risk-based assessments that integrate hazard, use, and exposure information about a substance.† The CSIA's safety standard is "that no unreasonable risk of harm to human health or the environment will result from exposure to a chemical substance" (see footnote †). The act would provide US EPA with authority to enact a range of risk management measures if it finds that a chemical does not meet the safety standard under its intended conditions of use, including labeling requirements, restrictions on production or distribution quantity or uses, and bans and phase-outs of chemicals or uses.‡

The CSIA is but one of many TSCA reform bills that have been proposed in both the Senate and the House of Representatives. Moreover, it is unlikely that the CSIA would pass in the form of S. 1009. It is nevertheless useful to consider the CSIA to better understand how the PBT concept might be incorporated into TSCA reform legislation as legislators have shown keen interest in PBTs and how they are being addressed under TSCA (e.g., US Congress 2010).

Overall, the CSIA would establish a regulatory program not unlike the Canadian CEPA and CMP, whereby authorities evaluate the existing inventory of chemicals in commerce, prioritize chemicals for tiered risk assessment, and then apply tailored risk management if necessary. In that light, the CSIA utilizes persistence, bioaccumulation, and toxicity criteria in a manner very similar to the CMP: PBT determinations guide prioritization and safety assessments, but they do not trigger automatic risk management (as occurs in Japan's CSCL and, to a degree, under REACH as well). As in CEPA and TSCA, the CSIA also does not compel US EPA to make formal PBT determinations, meaning that US EPA is authorized to

---

* Chemical Safety Improvement Act, http://cen.acs.org/content/dam/cen/91/web/S-1009
-113th-Congress.pdf.
† CSIA, § 3(4).
‡ CSIA, § 6(c)(9).

integrate persistence and bioaccumulation criteria into prioritization and assessments without applying formal PBT labels to lists of chemicals.

In fact, the terms *persistence* and *bioaccumulation* appear only once throughout the entire CSIA. US EPA is authorized to issue orders, rules, or formal agreements to obtain new data during the safety assessment and determination phases.* US EPA may issue guidelines for industry to follow when responding to orders or requests for new test data. CSIA § 4(j)(2) lists the types of health and environmental information for which the US EPA could issue guidelines, and persistence, bioaccumulation, and toxicity are among those types (as well as biomonitoring data and data about aggregate effects).† Although this is the only explicit reference to persistence and bioaccumulation in the draft bill, persistence and bioaccumulation would likely play a role in safety assessment and determination, given that this subsection indicates that they are relevant "health and environmental information" (see footnote *).

The CSIA tasks US EPA with developing its own method for conducting safety assessments, which must include a weight of evidence approach,‡ and also requires safety assessments to evaluate hazard, use, and exposure data.§ To reach a safety determination, US EPA must also use a weight of evidence approach to evaluate risk and consider "the magnitude of the risk posed by the chemical substance under the intended conditions of use..."¶ For both safety assessments and determinations, US EPA must consider reasonably available data from estimation methods (e.g., Quantitative Structure–Activity Relationships) and read-across methods that are "relevant to [estimating or] predicting the potential the environmental and human health effects, environmental and biological fate and behavior, and exposure potential for the substance."** To the extent that persistence and bioaccumulation are relevant to predicting the environmental and biological behavior of a substance, US EPA must consider them in safety assessments and determinations. Even if these provisions do not compel US EPA to consider persistence, bioaccumulation, and toxicity criteria, the agency is clearly permitted to consider the properties in each of these processes.

Persistence and bioaccumulation would also likely play a role in the prioritization process. The CSIA tasks US EPA with establishing criteria for determining whether substances are high- or low-priority substances for safety assessment and determination.†† US EPA must identify substances

---

* CSIA, § 6(b)(5)(B) (safety assessments); § 6(c)(8)(B) (safety determinations).
† CSIA, § 4(j)(2)(A)(ii).
‡ CSIA, § 6(b)(4)(A).
§ CSIA, § 6(b)(4)(D).
¶ CSIA, § 6(c)(3).
** CSIA, § 4(c)(3), (4).
†† CSIA, § 4(e)(2)(A).

that have the potential for high hazard and high exposure, leading to a category of high-priority substances.* Substances that have the potential for either high hazard or high exposure may also be labeled high priority. In addition to requiring US EPA to consider the data described above (available data from estimation and read-across methods), the CSIA also requires US EPA to consider "the hazard and exposure potential of the chemical substance...including specific scientific classifications and designations by authoritative governmental entities."† Persistence, bioaccumulation, and toxicity are relevant to determining hazard and exposure potential, and this subsection also seems to compel US EPA to consider formal PBT, vPvB, and POP designations (as well as others) by other authorities (e.g., states, the parties to the Stockholm Convention, and the European Commission). Given that the CSIA does not define "high hazard" or "high exposure," it seems reasonable to believe that US EPA may appropriate a version of the ranking system that it uses under TSCA—which integrates information on hazard (human and environmental toxicity), exposure, and persistence/bioaccumulation—to prioritize chemicals for safety assessments under the CSIA.

In our view, it would be wise for TSCA reform legislation to adhere to the risk management principles that are outlined in Chapter 1. In accordance with the principles of priority setting and exposure and risk assessment, a revised TSCA should incorporate US EPA's current prioritization system rather than compel US EPA to make formal PBT determinations. Although TSCA is not without its discontents, the manner in which it incorporates the PBT concept into its assessment and management scheme—through which US EPA considers persistence, bioaccumulation, and toxicity without making formal PBT identifications—is a positive feature worth replicating. As in the CSIA, US EPA should consider persistence, bioaccumulation, and toxicity alongside other important factors (e.g., production volume and uses) in prioritizing existing chemicals for risk assessment and management decisions. US policy makers could also learn from the example set by CEPA and the CMP to avoid compelling resource-intensive comprehensive risk assessments, which may not be necessary for US EPA to reach reasonable management decisions with a high level of confidence. Policy makers should also be mindful of what informational inputs are required at certain points in the assessment and management decision processes. Even though regulators might not be compelled to conduct full risk assessments, regulations should provide incentives for industry to generate high-quality data that can reduce reliance on estimation techniques. Finally, to the extent that a revised TSCA does incorporate formal PBT designations, such a designation should

* CSIA, § 4(e)(3)(E).
† CSIA, § 4(e)(2)(C).

not dictate any particular risk management outcome. Rather, legislation should provide for flexibility in the application of risk management options that may be tailored to particular uses, perhaps along the lines of the Canadian approach.

VI.   Conclusion

This chapter considers national PBT policies in Japan, China, Canada, and the United States. Chapter 4 considered the EU's REACH regulation, which is comparable in many ways with these national regimes. Each of the three types of PBT policies—priority setting, informational requirements, and risk management obligations—is present to some extent in each jurisdiction's chemicals policies. There is also variability among jurisdictions in the manner by which they utilize the PBT concept in their chemical regulations.

The existing chemical provisions under CEPA constitute the clearest example of the application of PBT characteristics to identify priority chemicals of concern. Persistence, bioaccumulation, and toxicity are explicitly mentioned together in CEPA's legislative text. PBT characteristics are considered alongside other factors (e.g., production volume) to identify chemicals of concern for tiered levels of risk assessment. The SLRAs conducted by Canadian authorities and the formal separation of risk assessment and risk management decisions under CEPA demonstrate how PBT characteristics can be effectively considered in both processes.

The US TRI program is a clear example of the PBT concept applied to informational requirements. A formal PBT designation is used to distinguish chemicals of highest concern, thereby triggering a lowering of the thresholds for general public information reporting requirements on emissions. The PBT concept is also a trigger for additional data production and/or submission requirements in the EU (e.g., additional data requirements under registration), Japan (e.g., information obligations for Monitoring Chemical Substances), and China (to the extent that PBTs are listed under Decree No. 591 or MEP Order No. 22).

One of the sharpest differences in PBT policies between jurisdictions is the risk management implication associated with formal PBT designations. Neither CEPA nor TSCA requires authorities to formally identify existing chemicals as PBTs, and no particular risk management outcomes are triggered by PBT information. A Class I PBT designation under Japan's CSCL, on the other hand, is viewed as a de facto ban on a chemical's import, production, and use. This seems to be the most stringent example of an automatic risk management implication from a PBT determination.

The REACH regulation utilizes the PBT concept in all three ways. PBT characteristics are utilized to prioritize potential SVHCs for assessment. SVHC designations carry informational and risk management obligations. The EU's experience in restructuring its risk management decision-making process into the Risk Management Options hints at the difficulties that can arise from attaching risk management obligations to formal labels. There have been growing pains associated with the highly complex way in which REACH is structured, but some of the inflexibilities have been addressed through careful guidance and creative implementation choices. It is nonetheless too early to determine how successful REACH will be in addressing PBTs in ways that satisfy the management principles.

Examining the design of these jurisdictions' chemical regulations and their experiences with the applying the PBT concept yields many findings and recommendations for the further development and harmonization of chemical governance regimes. These findings and recommendations are outlined in Chapter 7. But first, Chapter 6 considers the application of the PBT concept at the subnational level.

References

Abelkop, A.D.K., and J.D. Graham. 2015. "Regulation of Chemical Risks: Lessons for Reform of the Toxic Substances Control Act from Canada and the European Union." *Pace Environmental Law Review* 32.

Applegate, J.S. 2008. "Synthesizing TSCA and REACH: Practical Principles for Chemical Regulation Reform." *Ecology Law Quarterly* 35:721–69.

Arputharajoo-Middleton, R. 2014. "Companies Await More Details on China's Priority Hazchems." *Chemical Watch*, June 26.

Auer, C., and J. Alter. 2007. "The Management of Industrial Chemicals in the USA." In *Risk Assessment of Chemicals: An Introduction*, edited by C.J. van Leeuwen, and T.G. Vermeire, 553–74. Dordrecht, The Netherlands: Springer.

Berger, T.C., and M.E. Marrapese. 2004. "Notification of New Chemical Substances in Japan." Washington, DC: Keller & Heckman LLP, March. Available at http://www.khlaw.com/1117.

Briand, A. 2010. "Reverse Onus: An Effective and Efficient Risk Management Strategy for Chemical Regulation." *Canadian Public Administration* 53:489–508.

Chemical Watch. 2011a. "Japan Expects 10,000 Substances to be Notified by First Deadline." *Chemical Watch*, June 28.

Chemical Watch. 2011b. "Law Reforms Begin to Bite in Japan, Korea and China." *Chemical Watch*, November 7.

Chemical Watch. 2012. "Japan Releases over 7,000 Substance Screening Assessments." *Chemical Watch*, August 15.

Chemical Watch. 2013. "Japan Bans HBCD and Endosulfan." *Chemical Watch*, July 8.

ChemSec. 2013. "China Identifies 58 Chemicals to Act on," January–March. Available at http://www.chemsec.org/news/news-2013/january-march/1135-china-identifies-58-chemicals-to-act-on.

Chen, Y., J. Li, L. Liu, and N. Zhao. 2012. "Polybrominated Diphenyl Ethers Fate in China: A Review with an Emphasis on Environmental Contamination Levels, Human Exposure and Regulation." *Journal of Environmental Management* 113:22–30.

Chynoweth, E. 2014. "MEP Considering Changes to Order No. 22 Implementation Rules." *Chemical Watch*, November 7. Available at https://chemicalwatch.com/asiahub/21843.

CIRS (Chemical Inspection and Regulation Service). 2013. "Chemical Regulatory Report for China—Q1 2013." Prepared for Nickel Institute, May 10. Available at http://www.nickelinstitute.org/~/media/Files/MembersArea/Emerging Issues/ChinaRegulatoryUpdate/ChinaRegulatoryUpdateQ12013.ashx.

CIRS (Chemical Inspection and Regulation Service). 2014. "The Measures for Environmental Administration Registration of Hazardous Chemicals (2012)." Last modified April 10, 2014. Available at http://www.cirs-reach.com/China_Chemical_Regulation/The_Measures_for_Environmental_Administration_Registration_of_Hazardous_Chemicals_MEP_Order_22.html.

CSM (Chemical Sensitivities Manitoba) and CELA (Canadian Environmental Law Association). 2009. "A Response to the Proposed Risk Management Approach for Chemicals Management Plan Industry Challenge Batch 3 Substances Published in *Canada Gazette* Part I, Vol. 143, No. 10—March 7," May 1. Available at http://www.cela.ca/sites/cela.ca/files/652%20CMP%20batch%203.pdf.

Denison, R. 2007. *Not That Innocent: A Comparative Analysis of Canadian, European Union, and United States Policies on Industrial Chemicals*. Washington, DC: Environmental Defense. Available at http://www.edf.org/sites/default/files/6149_NotThatInnocent_Fullreport.pdf.

Du, S. 2014. "MEP's First Chemical 'Blacklist' Prioritizes Assessment of Acute Effects and Excludes Numerous EDCs." Greenpeace Commentary, ChemLinked, REACH24 Consulting Group, April 30. Available at https://chemlinked.com/expert-article/meps-first-chemical-blacklist-prioritizes-assessment-acute-effects-and-excludes-numerous-edcs.

Easton, S. 2008. "Canada's Chemicals Management Plan (CMP)." Presentation at the Meeting of the Great Lakes Binational Toxics Strategy Substance Working Group, Chicago, IL, April 8. Available at http://www.epa.gov/bns/integration/200804/Easton040808.pdf.

EC (Environment Canada). 2004. *A Guide to Understanding the Canadian Environmental Protection Act, 1999*. Ottawa, Quebec: Environment Canada.

EC (Environment Canada). 2008. *Final Screening Assessment Report for the Screening Assessment of [145 PBiT Substances]*, April. Available at http://www.ec.gc.ca/lcpe-cepa/documents/substances/pbti-pbit/final_145_PBiT-eng.pdf.

EC (Environment Canada). 2013a. "Search Engine for the Results of DSL Categorization." Last modified July 9. Available at http://www.ec.gc.ca/lcpe-cepa/default.asp?lang=En&n=5F213FA8-1&wsdoc=D031CB30-B31B-D54C-0E46-37E32D526A1F.

EC (Environment Canada). 2013b. "First Priority Substances List (PSL1)." Last modified June 21. Available at http://www.ec.gc.ca/ese-ees/default.asp?lang=En&n=95D719C5-1.

EC (Environment Canada). 2013c. "Second Priority Substances List (PSL2)." Last modified June 21. Available at http://www.ec.gc.ca/ese-ees/default.asp?lang=En&n=C04CA116-1.

EC (Environment Canada). 2013d. *Proposed Risk Management Measure for Polybrominated Diphenyl Ethers (PBDEs)*, February. Available at http://www.ec.gc.ca/ese-ees/92B7DD05-793A-4E4C-9742-3A25EB2529BE/PBDEs_Consultation_EN.pdf.

EC (Environment Canada). 2013e. "Polybrominated Diphenyl Ethers (PBDEs)." Last modified October 3. Available at http://www.ec.gc .ca/toxiques-toxics/Default.asp?lang=En&n=98E80CC61&xml=50464 70B-2D3C-48B4-9E46-735B7820A444.

EC (Environment Canada). 2013f. "Risk Management of DecaBDE: Commitment to Phase-Out Exports to Canada." Last modified July 23. Available at http:// www.ec.gc.ca/toxiques-toxics/default.asp?lang=en&n=F64D6E3B-1.

EC (Environment Canada). 2013g. "Polychlorinated Dibenzodioxins." Last modified July 23. Available at http://www.ec.gc.ca/toxiques-toxics/Default .asp?lang=En&n=98E80CC6-1&xml=1794091E-5FC5-40F9-BB0B -E823BFC418C6.

EC (Environment Canada). 2013h. "Polybrominated Biphenyls." Last modified August 8. Available at http://www.ec.gc.ca/toxiques-toxics/Default .asp?lang=En&n=98E80CC6-1&xml=7194BA9D-887F-4426-A2BE -E7E20560B67B.

EC (Environment Canada). 2013i. "Toxic Substances Management Policy." Last modified August 6. Available at http://www.ec.gc.ca/toxiques-toxics/default .asp?lang=En&n=2A55771E-1.

EC (Environment Canada). 2013j. "Track 1—Virtual Elimination from the Environment." Last modified July 23. Available at http://www.ec.gc.ca /toxiques-toxics/default.asp?lang=En&n=13698512-1.

EC (Environment Canada). 2013k. "LOQ Soil." Last modified August 14. Available at http://www.ec.gc.ca/toxiques-toxics/default.asp?lang=En&xml =61523358-F670-47A9-B34E-C1457A3EA5CB.

EC (Environment Canada). 2014. *Rapid Screening of Substances from Phase One of the Domestic Substances List Inventory Update: Results of the Final Screening Assessment*, March. Available at http://www.ec.gc.ca/ese-ees /7340E1B7-1809-4564-8C49-F05875D511CB/FSAR_RSII_EN.pdf.

EC (Environment Canada) and HC (Health Canada). 2011. *Proposed Risk Management Approach for Hexabromocyclododecane (HBCD)*, November. Available at http:// www.ec.gc.ca/ese-ees/5F5A32FB-3FD2-438F-A0A3-E973380199AF/HBCD _RM%20Approach_EN.pdf.

EC (Environment Canada) and HC (Health Canada). 2013. *Rapid Screening of Substances of Lower Concern: Results of the Screening Assessment*, April. Available at http://www.ec.gc.ca/ese-ees/2A7095CD-A88C-4E7E-B089 -486086C4CBC4/RSI%20Final%20-%20EN.pdf.

ECHA (European Chemicals Agency). 2015. "Grouping of Substances and Read-Across." Accessed January 10. Available at http://echa.europa.eu/en/support /grouping-of-substances-and-read-across.

Environmental Defence. 2011. *Canada's Chemical Management Plan: Progress Analysis 2006–2011*. Toronto, ON: Environmental Defence.

Fukushima, T. 2012. "Chemical Legislation and Regulations Updates: Japan." Director for Chemical Management Policy, Chemical Management Policy Division, Ministry of Economy, Trade, and Industry, Japan. Presentation at the Asia-Pacific Economic Cooperation Regulators' Forum, Singapore, March 30, 2012. Available at http://www.estis.net/includes/file.asp?site =cien-peru&file=82080D68-A181-4AFC-B4D0-6B94EBEB46AD.

Galatone, V. 2013. "Chemicals Management Plan: Moving Forward in 2013." Environment Canada. Presentation at the Industry Coordinating Group CEPA Update Conference, Mississauga, Ontario, June 6.

GAO (Government Accountability Office). 2013. *Toxic Substances: EPA Has Increased Efforts to Assess and Control Chemicals but Could Strengthen Its Approach*. GAO-13-249. Washington, DC, March 22. Available at http://www.gao.gov/products /GAO-13-249.

Government of Canada. 2011. "The Government of Canada 'Challenge' for Chemical Substances That Are a High Priority for Action." Last modified July 28. Available at http://www.chemicalsubstanceschimiques.gc.ca/challenge -defi/index-eng.php.

Government of Canada. 2012. "The Significant New Activity (SNAc) Approach." Last modified September 10. Available at http://www.chemicalsubstanc eschimiques.gc.ca/plan/approach-approche/snac-nac-eng.php.

Government of Canada. 2014a. "The Petroleum Sector Stream Approach." Last modified September 5. Available at http://www.chemicalsubstanceschi miques.gc.ca/petrole/index-eng.php.

Government of Canada. 2014b. "The Substance Groupings Initiative." Last modified December 4. Available at http://www.chemicalsubstanceschimiques .gc.ca/group/index-eng.php.

Government of Japan. 2009. "Chemicals." In *National Reporting to the Eighteenth Session of the Commission on Sustainable Development, Theme-Specific Issues*, December. Available at http://www.un.org/esa/dsd/dsd_aofw_ni/ni _pdfs/NationalReports/japan/Full_text.pdf.

Hamilton, J.T. 2005. *Regulation through Revelation: The Origin, Politics, and Impacts of the Toxics Release Inventory Program*. New York: Cambridge University Press.

Harper, S. 2006. "Canada's New Government Improves Protection against Hazardous Chemicals." *Prime Minister of Canada*. New Release, December 8. Available at http://www.pm.gc.ca/eng/news/2006/12/08/canadas-new -government-improves-protection-against-hazardous-chemicals.

Hashizume, N., Y. Inoue, H. Murakami, H. Ozaki, A. Tanabe, Y. Suzuki, T. Yoshida, E. Kikushima, and T. Tsuji. 2013. "Resampling the Bioconcentration Factors Data from Japan's Chemical Substances Control Law Database to Simulate and Evaluate the Bioconcentration Factors Derived from Minimized Aqueous Exposure Tests." *Environmental Toxicology and Chemistry* 32:406–9.

Hirai, Y. 2012. "Chemical Risk Assessment under the Chemical Substances Control Law in Japan and comparison with REACH." Risk Analysis Division, Chemical Management Center, National Institute of Technology and Evaluation, Japan. Presentation at the Society for Environmental Toxicology and Chemistry: Sixth World Congress/Twenty-Second Europe Annual Meeting, Berlin, Germany, May 20–24, 2012. Available at http://www.safe .nite.go.jp/risk/archive_pdf/SETAC2012_specialsession_nite.pdf.

Hogue, C. 2013. "Support Grows for Chemical Law Reform." *Chemical & Engineering News* 91:22–3.

Hu, J., T. Zhu, and Q. Li. 2007. "Organochlorine Pesticides in China." In *Persistent Organic Pollutants in Asia: Sources, Distributions, Transport and Fate*, edited by A. Li, S. Tanabe, G. Jiang, J.P. Giesy, and P.K.S. Lam, 159–211. New York: Elsevier.

Ikemoto, T. 2011. "Japan's Efforts on Management of PFOS." Chemicals Evaluation Office, Environmental Health Department, Ministry of the Environment, Japan. Presentation for the Organization of Economic Cooperation and Development Webinar on Alternatives to Long Chain PFCs, April 18. Available at http://www.oecd.org/env/ehs/risk-management/47643243.pdf.

Joint Committee. 2008. "II. New System of the Act on the Evaluation of Chemical Substances and Regulation of Their Manufacture, etc. (the Chemical Substances Control Law) to be established by 2020." Abstract. Joint Committee to Review the Chemical Substances Control Law, December 22. Available at http://www.meti.go.jp/policy/chemical _management/english/cscl/files/publications/review/joint_committee _report_abstract.pdf.

Kexiong, D. 2014. "The Latest Policies for Chemical Environmental Management in China." Deputy Director Level Officer. Division of Chemicals Management, Department of Pollution Prevention and Control, Ministry of Environmental Protection, People's Republic of China. Presentation at the Open Seminar on Chemicals Management Policies among China, Japan and Korea, November 12, 2014. Available at http://www.chemical-net.info/pdf/20141113_Seminar1 _eng.pdf.

Kimura, M. 2013. "Latest Trends on the Enforcement of Chemical Substances Control Law in Japan." Chemicals Evaluation Office, Ministry of the Environment, Japan. Presentation at the International Open Seminar on Chemicals Management Policies among China, Japan and Korea, November 15, 2013. Available at http://www.chemical-net.info/pdf/D3-3-JAPAN -KIMURA-EN.pdf.

Kogan, L.A. 2013. "REACH Revisited: A Framework for Evaluating Whether a Non-Tariff Measure Has Matured into an Actionable Non-Tariff Barrier to Trade." *American University International Law Review* 28:489–668.

Lau, M.H.Y., K.M. Leung, S.W. Wong, H. Wang, and Z.G. Yan. 2012. "Environmental Policy, Legislation and Management of Persistent Organic Pollutants (POPs) in China." *Environmental Pollution* 165:182–92.

Li, J., Y. Lu, Y. Shi, T. Wang, G. Wang, W. Luo, W. Jiao, C. Chen, and F. Yan. 2011. "Environmental Pollution by Persistent Toxic Substances and Health Risk in an Industrial Area of China." *Journal of Environmental Sciences* 23:1359–67.

Lin, F. 2012. "Regulations for Environmental Management on the First Import of Chemicals and the Import and Export of Toxic Chemicals." ChemLinked, REACH24 Consulting Group, June 17. Available at https://chemlinked .com/regulatory-database/regulations-environmental-management-first -import-chemicals-and-import-and.

Lin, F. 2013. "HCPEC Horizons: Inferences Drawn from the 12th FYP and Prospective Compilation of the HCPEC List." ChemLinked, REACH24 Consulting Group, April 15. Available at https://chemlinked.com/news /chemical-news/hcpec-horizons-inferences-drawn-12th-fyp-and-prospec tive-compilation-hcpec-list#sthash.J8IE6AEW.dpbs.

Liu, L. 2013. "MEP Notice 20 of 2013: The 12th Five-Year Plan on Environmental Risk Control of Chemicals." ChemLinked, REACH24 Consulting Group, June 27. Available at https://chemlinked.com/expert-article/ebook/china-mep -notice-20-2013-12th-five-year-plan-environmental-risk-control-chemicals.

Liu, L. 2014. "Decree 591—Regulations on the Control over Safety of Hazardous Chemicals." ChemLinked, REACH24 Consulting Group, June 10. Available at https://chemlinked.com/chempedia/decree-591#sthash.Op5hKvMj.dpbs.

Liu, Z., H. Wang, P.L. Carmichael, E.J. Deag, R. Duarte-Davidson, H. Li, P. Howe et al. 2012. "China Begins to Position for Leadership on Responsible Risk-Based Global Chemicals Management." *Environmental Pollution* 165:170–3.

MacDonald, S., and S. Castellarin. 2013. "Understanding the Risk Management Process." Health Canada and Environment Canada. Presentation at the Industry Coordinating Group CEPA Update Conference, Mississauga, Ontario, June 4–6, 2013.

McKenzie, D. 2013. "In China, Cancer Villages a Reality of Life." *CNN*, May 29. Available at http://www.cnn.com/2013/05/28/world/asia/china-cancer-villages-mckenzie/.

Meek, M.E., and V.C. Armstrong. 2007. "The Assessment and Management of Industrial Chemicals in Canada." In *Risk Assessment of Chemicals: An Introduction*, edited by C.J. van Leeuwen, and T.G. Vermeire, 591–621. Dordrecht, The Netherlands: Springer.

METI (Ministry of Economy, Trade, and Industry). 2010. "Notification of the Manufacturing Amount, etc. of General Chemical Substances and Priority Assessment Chemical Substances: Preliminary Preparation Materials." Tokyo, Japan: Chemical Safety Office, Chemical Management Policy Division, Manufacturing Industries Bureau, Ministry of Economy, Trade and Industry, December. Available at http://www.meti.go.jp/policy/chemical_management/english/files/CSCL-setsumei-H22-12-jizen-12eng.pdf.

METI (Ministry of Economy, Trade, and Industry). 2011. "Chemical Substances Control Law." Tokyo, Japan: Chemical Safety Office, Chemical Management Policy Division, Manufacturing Industries Bureau, Ministry of Economy, Trade and Industry. Available at http://www.meti.go.jp/policy/chemical_management/english/cscl/files/about/about_points.pdf.

METI (Ministry of Economy, Trade, and Industry). 2015. "History of the Law." Accessed January 13, 2015. Available at http://www.meti.go.jp/english/information/data/chemical_substances01.html.

METI (Ministry of Economy, Trade, and Industry), MLHW (Ministry of Health, Labour, and Welfare), and MOE (Ministry of the Environment). 2009. "Amendment of the Chemical Substances Control Law," May. Available at http://www.meti.go.jp/policy/chemical_management/english/files/Amendment_of_CSCL_JPN.pdf.

Naiki, Y. 2010. "Assessing Policy Reach: Japan's Chemical Policy Reform in Response to the EU's REACH Regulation." *Journal of Environmental Law* 22:171–95.

Nakai, S., K. Takano, and S. Saito. 2006. "Bioconcentration Prediction under the Amended Chemical Substances Control Law of Japan." *Trans. Sumitomo Kagaku* 2006-I. Tokyo, Japan: Sumitomo Chemical Co., Environmental Health Science Laboratory. Available at http://www.sumitomo-chem.co.jp/english/rd/report/theses/docs/20060106_vpv.pdf.

Nanimoto, H. 2012. "Japanese Chemical Control Legislation (CSCL & ISHL) and GHS Implementation." Ministry of Economy, Trade, and Industry, Japan. Presentation at ChemCon The Americas, New Orleans, LA, December 3–7, 2012.

Ni, K., Y. Lu, T. Wang, Y. Shi, K. Kannan, L. Xu, Q. Li, and S. Liu. 2013. "Polybrominated Diphenyl Ethers (PBDEs) in China: Policies and Recommendations for Sound Management of Plastics from Electronic Wastes." *Journal of Environmental Management* 115:114–23.

Renn, O., and E.D. Elliott. 2011. "Chemicals." In *The Reality of Precaution: Comparing Risk Regulation in the United States and Europe*, edited by J.B. Wiener, M.D. Rogers, J.K. Hammitt, and P.H. Sand, 223–56. Washington, DC: Resources for the Future.

Rorije, E., E.M.J. Verbruggen, A. Hollander, T.P. Traas, and M.P.M. Janssen. 2011. "Identifying Potential POP and PBT Substances: Development of a New Persistence/Bioaccumulation-Score." RIVM Report 601356001/2011. Utrecht, The Netherlands: National Institute for Public Health and the Environment. Available at http://www.rivm.nl/bibliotheek/rapporten/601356001.pdf.

Scott, D.N. 2009. "Testing Toxicity: Proof and Precaution in Canada's Chemicals Management Plan." *Review of European Community & International Environmental Law* 18:59–76.

Scott, D.N. 2012. "Beyond BPA: We Need to Get Tough on Toxics." *Globe and Mail*, January 4.

Shengxian, M.Z. 2013. "Speech by MEP Minister Zhou Shengxian at 2013 National Work Meeting on Environmental Protection." Ministry of Environmental Protection, People's Republic of China, February 4. Available at http://english.mep.gov.cn/Ministers/Speeches/201303/t20130320_249648.htm.

Shi, Y. 2013. "Comparing Chemical Reporting Requirements in China, Korea and Japan." Chemical Inspection and Regulation Service. Presentation at the Third Annual Shanghai Summit on Meeting on Chemical Regulations in China, Korea and Japan, Shanghai, China, October 16–17, 2013. Available at http://www.cirs-reach.com/china_chemical_regulation/Comparing_Chemical_Reporting_in_China_Korea_Japan_Yunbo_Shi.pdf.

Shibata, Y., and T. Takasuga. 2007. "Persistent Organic Pollutants Monitoring Activities in Japan." In *Persistent Organic Pollutants in Asia: Sources, Distributions, Transport and Fate*, edited by A. Li, S. Tanabe, G. Jiang, J.P. Giesy, and P.K.S. Lam, 3–30. New York: Elsevier.

Shimizu, M. 2011. "Risk Assessment of Chemicals in Japan." Chemicals Evaluation Office, Environmental Health Department, Ministry of the Environment, Japan. Presentation at the Open Seminar on Chemicals Management Policies, September 1, 2011. Available at http://www.chemical-net.info/pdf/20110901_S2-2_eng.pdf.

Takahashi, R. 2014. "Latest Development of Chemical Substances Control Law in Japan." Chemical Evaluation Office, Ministry of Environment, Japan. Presentation at the Open Seminar on Chemicals Management Policies among China, Japan and Korea, November 13, 2014. Available at http://www.chemical-net.info/pdf/20141113_Seminar2_eng.pdf.

Thompson, M., W. Feng, C. Hu, J. Li, and J.S. Eldred. 2011. "State Council of China Announces Revised Regulation on Control Over Safety of Dangerous Chemicals." Washington, DC: Keller & Heckman LLP, April 19. Available at https://www.khlaw.com/4434.

Toda, E. 2007. "The Management of Industrial Chemicals in Japan." In *Risk Assessment of Chemicals: An Introduction*, edited by C.J. van Leeuwen, and T.G. Vermeire, 575–89. Dordrecht, The Netherlands: Springer.

Tremblay, J.F. 2013. "China Steps Up Toxin Controls." *Chemical & Engineering News* 91(9):10.

US Congress. House. Committee on Energy and Commerce. 2010. *The Toxic Substances Control Act and Persistent, Bioaccumulative, and Toxic Chemicals: Examining Domestic and International Actions: Hearing before the Subcommittee on Commerce, Trade, and Consumer Production. 111th Cong., 2nd Sess.*, March 4.

US EPA (US Environmental Protection Agency). 1997. *Waste Minimization Prioritization Tool, Beta Test Version. 1.0: User's Guide and System Documentation Draft.* EPA530–R–97–019. Washington, DC: US EPA, Office of Pollution Prevention and Toxics and Office of Solid Waste.

US EPA (US Environmental Protection Agency). 2005. *Resource Conservation Challenge (RCC): 2005 Action Plan. Version 1*, May. Available at http://webapp1.dlib.indiana.edu/virtual_disk_library/index.cgi/6825758/FID3536/pdfs/rcc-act-plan.pdf.

US EPA (US Environmental Protection Agency). 2011. *Discussion Guide: Background and Discussion Questions for Identifying Priority Chemicals for Review and Assessment*, August. Available at http://www.epa.gov/oppt/existingchemicals/pubs/Chem.Priorization.August2011.DiscussionGuideOnly.pdf.

US EPA (US Environmental Protection Agency). 2012a. "Priority Chemicals." Last modified November 15. Available at http://www.epa.gov/osw/hazard/wastemin/priority.htm.

US EPA (US Environmental Protection Agency). 2012b. *TSCA Work Plan Chemicals: Methods Document.* Washington, DC: Office of Pollution Prevention and Toxics, February. Available at http://www.epa.gov/oppt/existingchemicals/pubs/wpmethods.pdf.

US EPA (US Environmental Protection Agency). 2013a. "List of Chemicals for Assessment." Last modified April 22. Available at http://www.epa.gov/oppt/existingchemicals/pubs/assessment_chemicals_list.html.

US EPA (US Environmental Protection Agency). 2013b. "2013 EPA TSCA Work Plan and Action Plan Risk Assessments and Data Collection Activities Existing Chemicals." Last modified April 22. Available at http://www.epa.gov/oppt/existingchemicals/pubs/2013wpractivities.html.

US EPA (US Environmental Protection Agency). 2013c. "2010/2015 PFOA Stewardship Program." Last modified January 16. Available at http://www.epa.gov/oppt/pfoa/pubs/stewardship/.

US EPA (US Environmental Protection Agency). 2013d. "Perfluorooctanoic Acid (PFOA) and Fluorinated Telomers." Last modified November 4. Available at http://www.epa.gov/oppt/pfoa/index.html.

US EPA (US Environmental Protection Agency). 2014a. "Persistent Bioaccumulative Toxic (PBT) Chemicals Covered by the TRI Program." Last modified March 16. Available at http://www2.epa.gov/toxics-release-inventory-tri-program/persistent-bioaccumulative-toxic-pbt-chemicals-covered-tri.

US EPA (US Environmental Protection Agency). 2014b. *TSCA Work Plan for Chemical Assessments: 2014 Update.* Washington, DC: Office of Pollution Prevention and Toxics, October. Available at http://www.epa.gov/oppt/existingchemicals/pubs/TSCA_Work_Plan_Chemicals_2014_Update-final.pdf.

US EPA (US Environmental Protection Agency). 2015. "TSCA Work Plan Chemicals." Last modified January 8. Available at http://www.epa.gov/oppt/existingchemicals/pubs/workplans.html.

Uyesato, D., M. Weiss, J. Stepanyan, D. Park, K. Yuki, T. Ferris, and L. Bergkamp. 2013. "REACH's Impact in the Rest of the World." In *The European Union REACH Regulation for Chemicals: Law and Practice*, edited by L. Bergkamp, 335–70. New York: Oxford University Press.

Wang, H., Z.G. Yan, H. Li, N.Y. Yang, K.M. Leung, Y.Z. Wang, R.Z. Yu et al. 2012. "Progress of Environmental Management and Risk Assessment of Industrial Chemicals in China." *Environmental Pollution* 165:174–81.

Wei, D., T. Kameya, and K. Urano. 2007. "Environmental Management of Pesticidal POPs in China: Past, Present and Future." *Environment International* 33:894–902.

Williams, T. 2006. *Virtual Elimination of Pollution from Toxic Substances*. PRB 06-26E. Ottawa, ON, Canada: Canadian Parliamentary Information and Research Service. Available at http://www.parl.gc.ca/Content/LOP/researchpubli cations/prb0626-e.htm.

Wu, Y. 2012. "China's Big Step to Extend Controls to Existing Chemicals." *Chemical Watch*. January 2013.

Zhang, H., Y. Lu, Y. Shi, T. Wang, Y. Xing, and R.W. Dawson. 2005. "Legal Framework Related to Persistent Organic Pollutants (POPs) Management in China." *Environmental Science & Policy* 8:153–60.

Zhang, Q., Y. Wang, A. Li, and G. Jiang. 2007. "Polychlorinated Dibenzo-*p*-Dioxins, Dibenzofurans, and Biphenyls, and Polybrominated Diphenyls Ethers in China." In *Persistent Organic Pollutants in Asia: Sources, Distributions, Transport and Fate*, edited by A. Li, S. Tanabe, G. Jiang, J.P. Giesy, and P.K.S. Lam, 213–36. New York: Elsevier.

Zhou, J. 2014. "Big Cuts in China's Final HCPEC List." ChemLinked, REACH24 Consulting Group, April 11. Available at https://chemlinked.com/news /chemical-news/big-cuts-chinas-final-hcpec-list.

chapter six

Subnational and private sector PBT policies

I. Introduction

Much of the discussion on chemical regulation tends to focus on formal policy making at the international and national levels, but a substantial amount of chemical policy making and implementation, including many that target PBTs, has been undertaken at the subnational level and by private companies and organizations. There is also a high degree of variation in policy approaches taken at the subnational level and by private actors, perhaps even more policy experimentation than occurs at regional and international levels of policy making. As with previous chapters, our examination of subnational and private approaches is not meant as an exhaustive documentation of such policies. In reviewing a selection of these PBT policies, this chapter provides only a sampling of the approaches taken by subnational and private actors. We have chosen examples to illustrate particular differences and commonalities among applications of the PBT concept.

Subnational policies often share many of the attributes of policies enacted by national governments. The typology outlined in Chapter 1 identifies the three broad types of PBT policies: government-initiated prioritization schemes, the use of the PBT category as a trigger for mandatory generation or provision of information, and direct risk management through various types of restrictions on the use, manufacture, and importation of PBTs. Elements of these approaches are present in the subnational policies we discuss, but subnational policies also constitute a distinct category owing to some unique characteristics of policy making at the subnational level.

In the second section of this chapter, we identify a distinct set of PBT policies that are designed to encourage voluntary risk management by private actors throughout the supply chain. These include government-initiated policies that encourage voluntary risk management by industry, such as Environmental Protection Agency's (US EPA's) Design for the Environment (DfE) program. Other private sector policies take the form of chemical evaluation tools for use by chemical manufacturers, formulators, product manufacturers, retailers, or even members of the public. The tools help decision makers make choices on which chemicals to use or avoid in production and purchasing. We discuss the PBT Profiler developed by US EPA and the GreenScreen for Safer Chemicals developed by the nongovernmental organization (NGO) Clean Production Action. Finally, some policies take the form of information aimed at informing governmental decisions and/or market transactions. As an example, we discuss the "Substitute It Now" (SIN) List developed by the NGO International Chemical Secretariat (ChemSec).

Subnational and voluntary programs, as well as some programs at the national and international levels, often operate together to influence the selection of chemicals in the marketplace. We illustrate the influence of multiple programs with a concluding case study on polybrominated diphenyl ether (PBDE) flame retardants. With the case study, we also describe the effects of "retail regulation"—the deselection of chemicals of concern by consumer product retailers rather than by government or by chemical manufacturers themselves. In particular, the retail chains Wal-Mart and Target are both, in effect, regulating the chemical marketplace and advancing green chemistry by discouraging the use of certain chemicals of concern in consumer products (e.g., Everts 2010; Cutting et al. 2011; Bomgardner 2014). The private policy decisions by retailers are, in part, internal business decisions, but they may be encouraged by government policy making. In the context of PBDEs, much of that policy making has occurred at the subnational level. We thus begin this chapter by discussing subnational PBT policies.

II.   Subnational PBT policies

In the United States, there is a trend of increasing regulatory action on chemicals at the state level (Schifano, Tickner, and Torrie 2009; Hogue 2013). The public interest organization Safer States has identified 169 chemical risk management policies enacted across 35 states as of January 2015. In 2014 alone, 120 bills were proposed at the state level to regulate chemicals (Safer States 2015). Examining a broader spectrum of policies, the Lowell Center for Sustainable Production at the University of Massachusetts Lowell developed a database, now maintained by the Interstate Chemicals Clearinghouse (IC2), that documents more than 1500 state and local chemical policies that have been proposed or enacted from 1990 through the present (IC2 2015; LCSP 2015).* More than 500 have been enacted, although only a small fraction of these policies address PBTs.†

---

* The database includes both legislative and executive actions at all levels of local and state government in the United States. The database also includes a broader range of policies than those that address industrial chemicals, including, for example, policies on hazardous waste or precautionary decision making.
† The database indicates that only 11 policies explicitly target PBTs as a category. However, many others target individual PBTs (e.g., PBDEs), and many more utilize the PBT concept in a way that does not lead the database to identify them as PBT policies (e.g., as one prioritization factor among many others).

Subnational policies work alongside and within national and international frameworks, such as the Stockholm Convention on Persistent Organic Pollutants (e.g., CIEL 2005; Ditz 2007). Many sources report that increasing state-level regulatory activity is a response to historically constrained or inadequate regulation under Toxic Substances Control Act (TSCA) and other statutes at the national level (e.g., Ditz 2007; Schifano, Tickner, and Torrie 2009; Allen 2013; Hogue 2013, 2014).

Many subnational policies are also enacted to protect particular regional or local natural resources or to reduce the exposure of local populations to particular chemicals of concern. For example, the province of Ontario and several US states have enacted policies to reduce environmental damage to the Great Lakes. The development of Washington State's PBT policies was, in part, motivated to address the effects of polychlorinated biphenyls (PCBs) and other chemicals in Puget Sound (CIEL 2005). California's policies on PBDEs were, in part, a response to the detection of high concentrations of brominated flame retardants in San Francisco Bay and in humans living in the region (CIEL 2005).

Literature on federalism and policy making at the state level often declares that states are "laboratories of democracy." In his famous dissenting opinion in *New State Ice Co. v. Liebmann* (1932), Supreme Court Justice Louis Brandeis observed: "It is one of the happy incidents of the federal system that a single courageous State may, if its citizens choose, serve as a laboratory; and try novel social and economic experiments without risk to the rest of the country."* Subnational policy experimentation tends to produce a high degree of variation in policy approaches, and chemical policies are no exception.

In the context of chemicals, state policies target particular products (e.g., children's products), individual chemicals or chemical groups, and categories of chemicals such as PBTs (Easthope and Valeriano 2007). Approaches include pollution prevention, data collection, environmental and public health monitoring, public information disclosure, risk assessment requirements, government purchasing standards, labeling requirements, waste disposal standards, and restrictions on use, manufacture, and importation (Schifano, Tickner, and Torrie 2009; IC2 2015). New York City and Seattle, Washington, among other cities, have enacted preferential purchasing policies that discourage the city government from purchasing products containing PBTs.† A sizeable proportion of subnational

---

* Louis Brandeis, New State Ice Co. v. Liebmann, 285 U.S. 262, 311 (1932) (dissenting).
† A Local Law to Amend the Administrative Code of the City of New York, in Relation to Environmental Purchasing and the Establishment of a Director of Citywide Environmental Purchasing, City of New York, Local Law No. 118 (Dec. 29, 2005); A Resolution Relating to Persistent, Bioaccumulative, Toxic Chemicals (PBTs), Stating the City of Seattle's Intent to Reduce its Use of PBTs, and Setting Forth a Work Program, City of Seattle, Res. No. 30487 (July 1, 2002).

chemical policies tend to target individual (or closely related) chemicals (e.g., mercury, PCBs, PBDEs); however, there appears to be a trend toward regulations that address categories of chemicals, including PBTs (Schifano, Tickner, and Torrie 2009). In the following subsections, we discuss the use of the PBT concept in select policies enacted by the province of Ontario and the US states California, Washington, and Oregon.

A. Ontario, Canada

The Canadian province of Ontario regulates chemicals through its Toxics Reduction Act (TRA) of 2009, administered by the Ontario Ministry of the Environment (MOE).* Operators of facilities that use or manufacture identified "toxic substances" must track the uses and quantity of the chemicals and prepare plans to address the possibility of phasing out manufacture and use of the chemicals.† In addition, operators must submit annual public reports that provide data on quantities, uses, and control measures taken.‡ The legislation requires that such plans be prepared, but the act does not impose any restriction on manufacture or use; risk management is voluntary (MOE 2012). For the purposes of the act, toxic substances are those that appear in Canada's National Pollutant Release Inventory (NPRI), which requires firms to report releases of pollutants, analogous to the Toxic Release Inventory (TRI) in the United States.§

Ontario has used the PBT concept in two ways. The province first used it as one prioritization factor among several to arrive at an initial set of 47 priority chemicals to regulate from the 165 that were listed in the 2006 NPRI (MOE 2008, 2009). Since the NPRI provided exposure data, Ontario was able to use a PBT scoring system similar to the US EPA's WMPT to help it determine which of the 165 NPRI substances were worthy of focus.

The system is known as the Scoring and Ranking Assessment Model (SCRAM). SCRAM was developed by the Michigan Department of Environmental Quality's Surface Water Control Division and Michigan State University's National Food Safety and Toxicology Center and was initially used to identify chemicals of highest concern for the agreements on the North American Great Lakes Basin (MOE 2009). SCRAM incorporates uncertainty about input values into its scores, making it particularly useful as a scoring tool for chemicals for which there is a lack of data and for which regulators rely on modeling and estimation techniques (e.g., Quantitative Structure–Activity Relationships) (Snyder et al. 2000).

---

* Toxics Reduction Act (TRA), 2009, S.O. 2009, c. 19; Ontario Regulation 455/09.
† TRA, §§ 3–10.
‡ TRA, §§ 8, 10.
§ TRA, § 2; Ontario Regulation 455/09, § 3.

Snyder et al. (2000) and Mitchell et al. (2002) provide an explanation of how the model functions. SCRAM is usable only for individual chemicals and not mixtures. It is meant to prioritize chemicals for risk assessment by giving chemicals scores that can be ranked relative to one another. The scoring system considers information on persistence, bioaccumulation, and toxicity and gives each characteristic a score of between 1 and 5. Bioaccumulation is determined from bioaccumulation factor (BAF), bioconcentration factor (BCF), or octanol–water partition coefficient values. A value greater than 100,000 (for any of the criteria) is scored as 5. A value of 100 or less is scored as 1, and there are specified ranges for scores of 2 through 4. In addition to assigning a score for the bioaccumulation values, SCRAM also assigns an uncertainty score based on the type of data that is used. For example, a measured BAF gives an uncertainty score of 0, whereas an estimated BCF receives an uncertainty score of 5.

The measured half-life of a substance can vary among media. Therefore, a score is required for each of five media including biota, air, soil, sediment, and water. If one of the input values is unavailable, it is estimated by using multimedia models. Interestingly, the system includes half-life in biota, which is often not considered in regulatory persistence determinations. The system uses the highest score of the five. SCRAM then gives an uncertainty score for the data, giving a score of 1 for estimation and a score of 2 for each category that does not have a value.

The persistence and bioaccumulation scores are then multiplied together to emphasize these properties relative to toxicity and to use them as an exposure score. An additional (arbitrary) weight of 1.5 is applied to the product of persistence and bioaccumulation to increase the influence of exposure and fate in the overall score relative to hazard. Toxicity is then considered—acute toxicity if the persistence score is 1 or 2 or chronic toxicity if the persistence score is 3, 4, or 5.

Acute toxicity is determined for terrestrial and aquatic toxicity individually. For acute terrestrial toxicity, ED_{50} (effective dose for 50% of the test population) or LD_{50} (lethal dose for 50% of the test population) are examined in five different media including plants, invertebrates, reptiles/amphibians, birds, and mammals. Each subcategory is assigned a score between 1 and 5, with the highest chosen. If there are no data for one of the subcategories, then it is given an uncertainty point. Aquatic toxicity applies the same process by examining EC_{50} or LC_{50} values in plants, amphibians, warm-water fish, cold-water fish, and invertebrates. Once these scores are tabulated, the final chemical score is determined.

For chronic toxicity, users provide data on subchronic/chronic toxicity for terrestrial organisms, aquatic organisms, and humans. To assess chronic toxicity for these different types of organisms, SCRAM calls for various types of data across different media, including Lowest and No Observed Adverse Effect Level and Lowest and No Observed Effect

Concentration data. The system assigns a severity factor, which gives more weight to severe effects than to moderate effects. For human chronic toxicity, the system calls for data on a variety of endpoints including carcinogenicity (which is treated with a weight of evidence factor) and general, reproductive, and developmental toxicity. The final score is then calculated using the scores for bioaccumulation, persistence, and the highest of the chronic toxicity scores.

Adding the final chemical score to the final uncertainty score generates the final composite score. This composite score is then used to generate a ranking of chemicals that accounts for missing data and the uncertainty in measurements. Therefore, if a chemical has estimated low values but high uncertainty, it is ranked higher than it would be if ranking were done without accounting for uncertainty.

Using this system (coupled with another program called the Risk Screening Environmental Indicators model) and the NPRI emissions data, the Ontario MOE determined to focus its risk management efforts on 47 of the 165 2006 NPRI substances after several stages of review (MOE 2009). This process provides another example of how a scoring system has been used to identify priority chemicals for risk management.

Ontario's regulation also employs the PBT concept to determine "substances of concern"—chemicals that are not tracked by the NPRI.* At present, the act lists 19 such substances (MOE 2009). To arrive at these 19, the MOE considered lists of CMRs, including California's Proposition 65, a list of 600 potential PBTs in the Great Lakes Basin listed in an academic report (Muir and Howard 2006), and lists of PBTs, including chemicals targeted by the Chemicals Management Plan (CMP) (MOE 2009).

Overall, Ontario's program reflects general characteristics of PBT policies at the subnational level, including constraints on resources, working within a national policy framework, and emphasizing the protection of local natural resources. First, its risk management approach utilizes informational requirements to encourage voluntary risk management by industry rather than directly restricting production or use.

Second, the program operates within the policy context set by national legislation. Ontario's regulation formally relies on regulatory determinations made in Canada's national NPRI program. The chemicals that it subjects to risk management are primarily defined by the NPRI, not by PBT characteristics. PBT characteristics played a role in determining initial priority chemicals to identify as "toxic substances," adhering to the priority-setting principle. A recently proposed Living List Framework for maintaining and updating the two lists also utilizes PBT characteristics in addition to considering uses and releases in Ontario (MOE 2014). Thus,

---

* TRA, § 11.

Ontario has adhered to the differentiation of uses and exposure and risk assessment principles.

Third, the MOE also consulted other authoritative lists of PBTs to identify additional chemicals to subject to reporting requirements. The selection of "substances of concern" is an example of one regulatory authority formally utilizing PBT determinations made by authorities in other jurisdictions—a feature that is common in several of the other subnational chemical regulations discussed in this chapter and that might be gaining traction in the implementation of some national programs as well.* Formal recognition of regulatory determinations made by authorities in other jurisdictions is one application of the value of information principle: It preserves public resources rather than generating a new regulatory list or conducting a new risk assessment.

Finally, Ontario's program is highly tied to the overall preservation of environmental quality in a particular geographic region—the Great Lakes Basin. The SCRAM was initially developed to identify priority chemicals of concern in the Great Lakes Basin. The TRA in Ontario is at a relatively early stage of implementation, and concerns have been expressed about its execution (Castrilli and de Leon 2014), but it nonetheless reflects strong public concern about PBTs.

B. *California, United States*

The state of California uses the PBT concept as a prioritization tool in its Safer Consumer Products (SCP) regulation,[†] which is designed to implement the legislative mandates of the state's Green Chemistry Initiative.[‡] As its name suggests, the SCP regulation addresses chemical safety by compelling companies to replace or reduce the use of hazardous chemicals at the product design phase.

---

* For example, in conducting rapid screening assessments under the CMP, Canadian authorities consult priority and regulatory lists in other jurisdictions, and US EPA consults other jurisdictions' PBT lists in identifying candidate chemicals for the chemical Work Plan. Also, § 4(e)(2)(C)(ii) of the proposed US Chemical Safety Improvement Act indicates that US EPA may consider "scientific classifications and designations by authoritative governmental entities."

† California Code of Regulations, Title 22, §§ 69501–69510 (2013) (hereinafter 22. Cal. Code Regs.), available at http://www.dtsc.ca.gov/LawsRegsPolicies/Regs/upload/SCP-Final -Regs-Text-10-01-2013.pdf.

‡ The Green Chemistry Initiative includes California Assembly Bill 1879 (grating Department of Toxic Substances Control authority to identify and assess chemicals of concern) and Senate Bill 509 (establishing a Toxics Information Clearinghouse to provide information to consumers).

The SCP regulation went into effect in October 2013 and includes four primary components: identification of Candidate Chemicals, identification of Priority Products, alternatives analysis (AA), and regulatory response.[*]

First, the regulation instructs the Department of Toxic Substances Control (DTSC) to identify and prioritize Candidate Chemicals.[†] Candidates include chemicals that exhibit a "hazard trait and/or an environmental or toxicological endpoint" and are included on at least one list from two groups of specified lists developed by the government of California or regulatory agencies in other jurisdictions.[‡] The first group includes 15[§] lists that identify chemicals with particular hazard traits, including CMRs, endocrine disruptors, neurotoxicants, respiratory sensitizers, and PBTs. This first group of lists includes chemicals identified as PBTs under the US TRI program, the US National Waste Minimization program, Washington State's PBT Strategy, PBTs and very persistent, very bioaccumulative (vPvB) chemicals identified as substances of very high concern (SVHCs) under the European Union (EU) Registration, Evaluation, Authorization, and Restriction of Chemicals (REACH) regulation, and potential PBiTs identified from the Canadian Domestic Substances List (DSL) categorization. The second group includes eight lists of chemical types that are associated with particular exposure indicators. This group includes, for example, priority chemicals under the California Environmental Contaminant Biomonitoring Program and designated chemicals for priority action under the Oslo–Paris Convention, both of which include PBTs.

In addition to incorporating the PBT concept into its identification of candidate chemicals through reference to these lists, the definition of "hazard trait" in the California Code of Regulations also includes those chemicals with PBT properties.[¶] And, in addition to including a standard definition of persistence and bioaccumulation (defined by half-life and BCF), California's definitions of persistence and bioaccumulation include those chemicals listed as persistent and bioaccumulative by another jurisdiction.[**]

Altogether, the authoritative lists include about 2900 chemicals. In September 2013, the DTSC listed 1060 chemicals on its initial Candidate Chemicals List and prioritized 164 candidate chemicals that appear on at

---

[*] For overviews of the SCP regulation, see DTSC (2013), Cowan et al. (2014), and Margulies and Troutman (2013).

[†] 22 Cal. Code Regs., § 69502.2.

[‡] 22 Cal. Code Regs., § 69502.2.

[§] The group includes 13 lists, but subdivides two of the lists to yield a total of 15.

[¶] 22 Cal. Code Regs., §§ 69501.1(18), (36); 69405.2, 69405.3.

[**] See, e.g., 22 Cal. Code Regs., § 69405.2(b) ("Evidence for the bioaccumulation hazard trait includes but is not limited to: the identification of a substance to be bioaccumulative by an authoritative organization…").

least one list from both groups of authoritative lists for hazard traits and exposure indicators (Cowan et al. 2014; DTSC 2015). The DTSC plans to review the list at least once per year (DTSC 2015). The agency may remove chemicals from the list and also add chemicals to the Candidate Chemicals List if, for example, a chemical presents a risk of "adverse impacts." The SCP regulation identifies specific factors to consider when determining if a chemical presents the risk of adverse impacts,[*] including a chemical's hazard traits (e.g., PBT properties) and its environmental fate, the definition of which also includes persistence and bioaccumulation.[†]

The second component of the SCP regulation is the identification of Priority Products, which denote chemical–product relationships— essentially, products that present exposure pathways for one or more candidate chemicals. In this process as well, the SCP regulation incorporates the PBT concept into its prioritization and identification process. Priority Products must present both a potential for exposure and a potential for "significant or widespread adverse impacts."[‡] The SCP regulation uses the PBT concept, as well as other characteristics, to identify "adverse impacts and exposures."[§] Here again, the SCP regulation allows the DTSC to consider PBT properties by reference to a chemical's hazard traits and environmental fate.

The DTSC may also consider a chemical's "physicochemical properties," which include (but are not limited to) a chemical's physical state; molecular weight; density; vapor pressure and saturated vapor pressure; melting point; boiling point; water solubility; lipid solubility; octanol-water partition coefficient, octanol-air partition coefficient, organic carbon partition coefficient; diffusivity in air and water; Henry's Law constant; sorption coefficient for soil and sediment; redox potential; photolysis rates; hydrolysis rates; dissociation constants; or reactivity including electrophilicity.[¶]

In other words, the DTSC may consider many of the factors that underlie a PBT determination without using the language of "PBT." The California Code of Regulations, like those in other jurisdictions, defines persistence and bioaccumulation with reference to specific cutoff values, but by authorizing DTSC to consider physicochemical properties without a reference to cutoff values, the SCP regulation builds flexibility into the DTSC's prioritization process, similar to what a weight of evidence approach might provide.

Thus, the SCP regulation incorporates the consideration of PBT properties into its prioritization of Candidate Chemicals and Priority Products

---

[*] 22 Cal. Code Regs., § 69503.3.
[†] 22 Cal. Code Regs., § 69501.1(32).
[‡] 22 Cal. Code Regs., § 69503.2(a)(2).
[§] 22 Cal. Code Regs., § 69503.2(b)(1)(A).
[¶] 22 Cal. Code Regs., § 69407.2.

through several overlapping mechanisms. Persistence, bioaccumulation, and toxicity are only three of many factors that DTSC may consider when it evaluates other jurisdictions' lists. The SCP regulation does not instruct DTSC to weigh PBT considerations more or less heavily than other considerations (e.g., concerns over CMRs, endocrine disruptors, and neurotoxins). There is no formulaic or numerical scoring system for determining priorities. Ultimately, structuring the prioritization process in this way provides DTSC with the flexibility to consider many factors—especially the physicochemical properties that underlie a PBT determination or exposure pathways—without forcing a decision based on numerical cut-off values.

The SCP regulation permitted DTSC to identify no more than five initial Priority Products. In March 2014, the agency released an initial list of three proposed Priority Products: paint and varnish strippers and surface cleaners containing methylene chloride, spray polyurethane foam systems containing unreacted diisocyanates, and children's foam-padded sleeping products containing tris(1,3-dichloro-2-propyl) phosphate (DTSC 2014a). None of these Priority Products raise PBT concerns. In September 2014, the DTSC released a draft Priority Product Work Plan, which explains its prioritization process for identifying Priority Products (DTSC 2014b). The regulation permits the DTSC to consider PBT properties in addition to a variety of other factors. In its draft Work Plan, the DTSC does not explicitly list PBT properties as an overriding concern. Rather, the DTSC identified prioritization factors as clear exposure pathways, biomonitoring studies, indoor air quality studies, impacts on sensitive subpopulations, impacts on aquatic resources, and water quality monitoring studies. These factors may lead the DTSC to identify Priority Products containing PBTs in the future, but the overriding prioritization factors seem to be driven by exposure-based concerns rather than hazard-based concerns, consistent with the exposure and risk assessment principle.

Once the DTSC identifies a chemical–product relationship by linking a Candidate Chemical to a Priority Product, the chemical becomes designated as a "Chemical of Concern."* The SCP regulation then places notification requirements on "responsible entities."† The regulation requires identification of all of the companies in a product's supply chain, including manufacturers, importers, assemblers, and retailers, as responsible entities. Responsible entities must notify the DTSC with information on their identity and a description of the product.‡

---

* 22 Cal. Code Regs., §§ 69501.1(21), 69503.5(b)(2)(B).
† 22 Cal. Code Regs., § 69501.1(60).
‡ 22 Cal. Code Regs., § 69503.7.

The third component of the SCP regulation is AA.* Generally, the SCP regulation requires responsible entities to conduct an AA for the Priority Product.† However, the SCP regulation gives responsible entities three other options.‡ They may remove the chemical from the product, remove the product from the California market, or replace the product–chemical combination with a reformulated product. The regulation specifies information notification requirements for each option. Moreover, Chemicals of Concern are exempted from AA if they are present in the product below an AA Threshold—a defined Practical Quantitation Limit below which the chemical cannot be reliably detected with routine laboratory practices.§

AA is conducted in two stages.¶ Firms that engage in AA must draft a preliminary AA, including information that identifies the requirements of the product, the functions of the Chemical of Concern, a list of alternative options, comparisons between the hazard profiles of alternative chemicals to that of the chemical of concern, and other factors including socioeconomic factors. Alternatives may include removal of a Chemical of Concern from the product, replacement with another chemical, redesign of the product to eliminate or reduce the concentration of the Chemical of Concern, redesign of the product or manufacturing process to reduce potential exposures to the Chemical of Concern, and any other change to the product or process to reduce potential adverse effects or exposures, including waste and end-of-life effects.**

If the DTSC approves the preliminary AA, then firms may draft a final AA, which involves a more thorough comparison of alternatives. The final AA must include a consideration of adverse health and environmental impacts, exposure, product function and performance, and socioeconomic impacts. The AA must also select a preferred alternative. The SCP regulation includes hazard traits, environmental fate, and physicochemical properties (and hence, PBT properties) as factors to consider in the comparison of alternative chemicals to Chemicals of Concern. The regulation lays out a framework by which DTSC must evaluate AAs. The AA process is intended to guide industry to make informed substitution decisions and to advance green chemistry. Although we are unable to comment on the AA process in practice as of yet, the design of the AA process appears consistent with the exposure and risk assessment, differentiation in uses, and rational alternatives principles (see Cowan et al. 2014).

---

* For an overview of the AA process, see Tickner et al. (2013), DTSC (2013), and Cowan et al. (2014).
† 22 Cal. Code Regs., § 69505.1.
‡ 22 Cal. Code Regs., § 69505.2.
§ 22 Cal. Code Regs., § 69501.1(52).
¶ 22 Cal. Code Regs., §§ 69505.4–7.
** 22 Cal. Code Regs., §§ 69501.1(10).

The final component of the SCP regulation is a regulatory response. Under specified conditions (e.g., if the AA does not select an alternative or if the DTSC disapproves of the AA), the SCP regulation permits the DTSC to enact a number of risk management options, including use restrictions, sales prohibitions, engineered safety measures, administrative controls, end-of-life management requirements, and requirements for research and development by industry.* Moreover, in selecting a regulatory response, the DTSC gives preference to the options with the "greatest level of inherent protection," meaning that the agency must generally prefer approaches that reduce the adverse impact of or exposure to a Chemical of Concern through a redesign of the product or production process rather than through an administrative or engineering control to reduce release of the chemical.†

The SCP regulation does not target PBTs alone, as shown by the DTSC's initial proposed Priority Products. Cowan et al. (2014) analyzed the authoritative lists and the DTSC's initial list of Candidate Chemicals to identify the most prevalent chemicals and therefore perhaps the most likely to be identified as Chemicals of Concern within Priority Products. They found that the most prevalent Candidate Chemicals tend to fall into either the PBT or CMR categories (although many of the most prevalent PBTs, such as dichlorodiphenyltrichloroethane [DDT] and other persistent organic pollutants [POPs], are already heavily regulated). The DTSC's draft Priority Product Work Plan identifies some possible product–chemical combinations that include PBTs, such as the use of brominated flame retardants in furniture, which might be identified as Priority Products in the future.

The primary use of the PBT concept within the SCP regulation is as a factor to identify and prioritize Candidate Chemicals and Priority Products. As such, the regulation's application of the PBT concept is consistent with the priority-setting principle. The overall design of the program also exemplifies the principles of exposure and risk assessment and differentiation in uses. The regulation focuses on exposure in addition to hazard by targeting chemical–product relationships, by prioritizing Candidate Chemicals that appear on both hazard trait and exposure indicator lists, and by prioritizing Priority Products based on exposure-related factors. Green chemistry emphasizes the development of safer alternative chemicals and the incorporation of those chemicals into products at the design phase. The SCP regulation therefore also fulfills the rational alternatives principle and, perhaps better than many other regulations, builds comparative risk analysis into the heart of the program. While the program is too new to evaluate, it does have a stronger focus on

---

* 22 Cal. Code Regs., §§ 69506.
† 22 Cal. Code Regs., §§ 69506(b).

the development of safer alternatives than do most, if not all, of the other programs we have discussed.

Finally, the SCP regulation also seems to be consistent with the value of information principle in its reliance on authoritative categorical determinations of other jurisdictions rather than compelling the DTSC to undergo a distinct PBT assessment. The regulation does not require informational inputs from industry until after it has identified Priority Products. The informational requirements of the AA process and the overall quality of the AAs, however, remain to be seen. The reliance on authoritative lists of other jurisdictions should conserve limited public resources.

Overall, the SCP is not a typical subnational policy. Although it is designed to address exposures to Chemicals of Concern within California (e.g., by emphasizing the identification of Priority Products by using biomonitoring and environmental surveillance from California), the SCP regulation will likely affect national, and possibly international, markets. There are potentially millions of product–chemical combinations associated with the 164 priority Candidate Chemicals (Cowan et al. 2014). By itself, California was the world's eighth largest economy in 2013 (CCSCE 2014). Moreover, the breadth of the regulation is much larger than that of other state regulations. In the following subsections, we discuss PBT policies in Washington State and Oregon, which are more limited in their application.

C.   Washington State, United States

Because of concerns over chemical pollution, especially the presence of mercury, dioxins, and PCBs in Puget Sound, in 2000, the Washington State legislature directed the state's Department of Ecology (DOE) to establish a strategy to address PBTs (DOE 2000). The DOE released its PBT Strategy later that year, establishing Washington State's basic approach to PBT policy: the use of PBT criteria to identify priority chemicals of concern followed by Chemical Action Plans (CAPs), which characterize hazard and risk and suggest different risk management options. The initial strategy listed mercury, PCBs, benzo(a)pyrene, dioxins and furans, DDT, aldrin/dieldrin, chlordane, hexachlorobenzene, and toxaphene as the priority chemicals to address.

The DOE focused its initial efforts on mercury and, in cooperation with the state Department of Health, published a CAP for mercury in 2003 (DOE and DOH 2003). A CAP "identifies, characterizes and evaluates uses and releases of a specific PBT…and recommends actions to protect human health or the environment."* A CAP does not include a full quantitative

---

* Washington Administrative Code, § 173-333 (2014) (hereinafter WAC), available online at http://apps.leg.wa.gov/wac/.

risk analysis, but it includes elements of a risk analysis (e.g., quantitative release and exposure estimates) as well as benefit–cost analyses of proposed risk management actions. Authorities include stakeholders in the process by developing CAPs in cooperation with affected industries. Finally, CAPs are only advisory recommendations; additional regulation or legislation is necessary to enact risk management. For example, in accordance with suggestions in the mercury CAP, in 2003, the Washington State legislature enacted the Mercury Education and Reduction Act.* The act establishes labeling requirements for certain products, prohibits the sale of many products that contain mercury, and compels the DOE to establish a disposal education plan for stakeholders.

After the execution of this process, the legislature and governor directed the DOE to develop a formal rule to address PBTs and begin work on a CAP for PBDEs. The DOE finalized its PBT rule in 2006.† The rule uses the PBT concept as a priority-setting tool to facilitate risk management. The precautionary goal of the rule is to "reduce and phase-out PBT uses, releases and exposures in Washington."‡ To that end, the rule lists 18 individual chemicals and eight chemical groups (e.g., perfluorooctane sulfonates, polycyclic aromatic hydrocarbons [PAHs], PBDEs, and PCBs).§ Listed chemicals must be persistent, bioaccumulative, and toxic. The rule also lists two metals—lead and cadmium.¶

Being on the list triggers no regulatory action itself, and the list is not the sole basis for any formal risk management decisions.** Rather, the DOE may use the list to establish CAPs, ambient monitoring, biomonitoring, public awareness, and voluntary risk management measures.†† CAPs remain the primary tool that the DOE uses to propose risk management options for listed chemicals. As of 2014, the DOE has finalized CAPs for mercury, PBDEs, lead, and PAHs (DOE 2014a). In accordance with the PBDE CAP, in 2007, the legislature enacted a law that prohibits the manufacture, sale, and distribution of products containing PBDE flame retardants, with limited exceptions.‡‡ The agency is presently developing a CAP for PCBs (DOE 2014b). The DOE maintains a multiyear schedule for the development of CAPs on its PBT list. Priorities are determined by an internal ranking system that considers PBT characteristics in addition to uses, releases, levels in the environment, and biomonitoring data, all in the state of Washington (DOE 2007).

---

* Washington Revised Code § 70.95M (2014). For a description, see CIEL (2005).
† WAC 173-133.
‡ WAC 173-333-100.
§ WAC 173-333-310(2).
¶ WAC 173-333-315(2).
** WAC 173-333-300(3)(b).
†† WAC 173-333-300(2).
‡‡ Washington Revised Code § 70.76 (2014).

As in Ontario and California, the design and implementation Washington State's PBT policy also reflect several of the general characteristics of PBT policies at the subnational level. The state's resources are focused on a limited number of chemicals that affect the environment and human health in the state of Washington. Unlike policies in Ontario and California, the DOE does not formally consider regulatory lists from other jurisdictions in identifying priorities.

Washington State's approach to PBTs includes elements of both hazard- and risk-based approaches. The overall goal to phase out PBTs (regardless of exposure level) is certainly hazard based and reflects the impression that the combination of persistence, bioaccumulation, and toxicity characteristics makes the complete control of risks virtually unattainable. It is a viewpoint similar to the premises that underpin the European treatment of PBTs as nonthreshold chemicals. In implementation, though, the state uses the PBT concept to establish priorities and assess uses, releases, and exposure—elements of a risk-based approach that also reflects the principle of differentiation of uses. Implementation is at a relatively early stage, as DOE has only released CAPs for four chemicals and chemical groups* and the legislature has only enacted risk management measures on three of them.[†]

D.   Oregon, United States

The state of Oregon has incorporated the PBT concept into several aspects of its overall Toxics Reduction Strategy and green chemistry initiative (Department of Environmental Quality [DEQ] 2012, 2014). The Oregon legislature has enacted laws to prohibit "[t]he introduction or delivery for introduction into commerce of any product containing more than" 0.1% by mass of the PBT flame retardants penta-, octa-, and deca-BDE.[‡] The focus of this subsection, however, is to highlight the state of Oregon's utilization of the partial PBT concept to define the scope of one of its laws.

In 2007, the state of Oregon's legislature enacted Senate Bill 737, which required Oregon's DEQ to develop a Priority Persistent Pollutant List by 2009 and report on exposures, sources, and management options by June 2010.[§] The statute defines "persistent pollutant" as "a substance that is toxic and *either* persists in the environment *or* accumulates in the tissues of humans, fish, wildlife, or plants."[¶] The Oregon legislature explicitly

---

* A draft CAP on PCBs was released in July 2014 (DOE 2014b).
† The state has enacted various controls on lead in addition to mercury and PBDEs.
‡ Oregon Revised Statutes, 453.085(16) (2013).
§ For information on Senate Bill 737, see Oregon Department of Environmental Quality, Water Quality, Senate Bill 737 at http://www.deq.state.or.us/wq/SB737/.
¶ Oregon Revised Statutes, Chapter 468B.138(4) (2013) (emphasis added).

decided to use the "or" construct to widen the scope of the legislation, thereby including some partial PBTs in addition to PBTs.

Hope et al. (2010) explain the process of establishing the list. Regulators considered lists and PBT criteria used in other state, national, and international regulatory programs. The authors state, "The primary consequence of an 'or' versus an 'and' construct is that the number of chemicals classified as PBTs or POPs can be substantially larger with an 'or' construction" (Hope et al. 2010, p. 735). Environment Canada also used an "or" construct for its DSL categorization and prioritization before the onset of the CMP, and the results (described in Chapter 5) corroborate this statement. The DEQ's list of Priority Persistent Pollutants includes 118 chemicals and chemical groups—69 persistent pollutants (PTs and BTs) as well as 49 "legacy pollutants," which have been prohibited but remain detectable in sediment and tissue (e.g., DDT, PCBs, and dioxins) (Parametrix 2008; DEQ 2009).*

Senate Bill 737 is concerned primarily with water quality. The policy's risk management of PBTs is limited to effluents from its 52 large municipal wastewater treatment plants. In 2010, the DEQ's Environmental Quality Commission issued a rule establishing threshold levels of persistent pollutants in the treatment plants' effluent streams.† If a facility exceeds the threshold, it must develop a pollution reduction plan. In 2011, the DEQ required several plants to develop pollution prevention plans for arsenic, beta-sitosterol, and pyrene, which were present above the specified levels (DEQ 2011; Keller and Heckman LLP 2014). Thus, the scope of this Oregon program is more limited than many other programs.

In Senate Bill 737, the state of Oregon uses PBT characteristics to identify priorities, but the use of the "or" construct to include partial PBTs widens the scope of the regulation to PTs that biota might excrete rather than accumulate and BTs that might break down in the environment relatively quickly. The PBT category exists because the three properties together pose unique concerns. That is not to say that particular partial PBTs do not raise concerns as well, or even greater concerns given the context. Rather, we are suggesting that there is merit in allocating limited public resources toward addressing the risks associated with those chemicals with all three characteristics. In this instance, the use of the "or" construct might make sense because the regulation applies only to a particular exposure pathway—effluent streams from municipal wastewater treatment plants. Different prioritization criteria might be desirable for regulations that require the dedication of more public resources to implement and enforce or for which CMR properties are greater concerns than aquatic toxicity.

* Oregon Revised Statutes, Chapter 468B.138(1) (2013), http://www.leg.state.or.us/ors/468b.html.
† Oregon Administrative Rule, Initiation Level Rule, 340-045-0100 (2013).

In summary, the uses of the PBT concept in the chemical policies of Ontario, California, Washington State, and Oregon illustrate a variety of applications. Whereas Oregon applies PBT characteristics to identify chemicals to control from the effluent streams in municipal wastewater treatment plants, California's SCP regulation may apply to any PBT–product combination. Whereas those regulations apply risk management to identified chemicals and exposure pathways through thresholds and AA requirements, respectively, the policies of Washington and Ontario facilitate the development of risk management plans to address any risks associated with individual PBTs rather than specific exposure pathways. In particular, Ontario's policy requires industry to develop risk management plans and to voluntarily abide by them. In the following section, we consider private sector PBT policies, including voluntary risk management programs, in greater detail.

III.   *Programs to encourage voluntary management of PBTs*

Government can encourage voluntary risk management by industry through nonregulatory approaches (Tickner et al. 2013). Indeed, one of US EPA's primary roles in the area of chemical risk management is as a "facilitator of stewardship" (Auer and Alter 2007). Moreover, the incentive for risk management could come from private actors rather than from the government.

The efficacy of voluntary risk management initiatives is mixed and contentious (e.g., Morgenstern and Pizer 2007; Gamper-Rabindran and Finger 2013). Nonetheless, risk management decisions by private actors are garnering significant attention in the area of chemical regulation, especially as it pertains to hazard categories like PBT (Easthope and Valeriano 2007; Greer 2010). Here, we call attention to four types of approaches that have been employed to encourage voluntary risk management of PBTs.

US EPA's DfE Program and PBT Profiler are examples of public approaches to provide private actors with information or evaluative tools that they can use to make risk management decisions. Clean Production Actions' Green Screen for Safer Chemicals and the International ChemSec's SIN List of chemicals of concern are examples of privately developed resources to facilitate voluntary risk management decisions.

Such approaches and tools are evaluative and/or informational in nature and can affect decisions throughout chemical supply chains. They provide information to chemical manufacturers making decisions on chemical design or inputs into chemical formulations, to product manufacturers choosing chemical inputs, to product wholesalers and retailers in choosing which products to sell, and finally, to consumers making choices about which products to purchase and use. Ultimately, the aim of these approaches in chemical risk management is to encourage private actors to voluntarily take action to reduce risk and to improve the quality of those actions.

A. Government programs to encourage voluntary risk management

1. Design for the Environment

US EPA's DfE program, administered by the Office of Pollution Prevention and Toxics, is intended to facilitate the voluntary substitution of hazardous chemicals with safer alternatives and to reduce the likelihood of unintended consequences that might result if poorly understood alternatives are chosen (US EPA 2014a).

Under the DfE program, US EPA works with industry, NGOs, and academics through partner projects (Auer and Alter 2007). DfE includes best practices, labeling, and AA programs. For example, DfE has established best practices guidelines to reduce worker and consumer exposure to chemicals of concern in automotive refinishing and in the use of spray polyurethane foam (US EPA 2014b). The labeling and alternatives assessment programs employ the PBT concept as a hazard criterion to grade chemicals.

As of 2013, about 2500 products (e.g., certain household cleaning products) have been marked with the DfE label as a safety and effectiveness signal to consumers (US EPA 2014c). To be marked with the DfE label, a product's ingredients must meet the Master Criteria for Safer Ingredients—"science-based criteria designed to ensure that the safest possible ingredients are used in DfE-labeled products" (US EPA 2012a, p. 2). The criteria call for evaluations of toxicological endpoints, including CMR properties, as well as environmental toxicity and fate—aquatic toxicity, persistence, and bioaccumulation.

DfE's labeling program evaluates the persistence, bioaccumulation, and toxicity properties conditionally and relative to one another. If a chemical is an acute aquatic toxicant, then it must have a short half-life to qualify for the label, and vice versa (US EPA 2012a). However, even substances with low aquatic toxicity must have half-lives of less than 60 days to qualify. All acute aquatic toxicants must also have a BCF/BAF of less than 1000. DfE applies even stricter criteria for products intended for direct release to the environment, such as graffiti removers and boat cleaners (US EPA 2012a, 2014d). These stringent requirements to qualify for the DfE label are designed to advance green and sustainable chemistry and encourage informed substitution.

Alternatives assessments use PBT information as well (US EPA 2011a). First, DfE participants determine whether partnership assessment projects will likely be feasible and worthwhile.

Chemical substitutes should be commercially available or at least likely to become available. Their use should be technically feasible and offer some prospective benefit in terms of performance or cost-effectiveness. Substitutes should also prospectively offer environmental or human health benefits relative to current chemicals. Stakeholders must show interest and willingness to participate. And, finally, participants consider whether alternatives are likely to result in practical change (Lavoie et al. 2010; US EPA 2014e).

Second, DfE participants will collect information on chemical substitutes, including information on how well characterized they are, manufacturing processes, and the like. After the scope of the project is defined, DfE draws stakeholder input from the entire supply chain, including chemical manufacturers, product manufacturers, retailers, consumers, waste handlers, NGOs, government agencies, and academics. Stakeholder consultations and reviews of the literature help identify alternatives that could be adopted with relatively minimal impact on the manufacturing process (US EPA 2014e).

Once the viable alternatives are identified, DfE calls for a hazard assessment of a variety of endpoints, including persistence, bioaccumulation, and toxicity. Unlike the labeling criteria, which products must meet to receive the label, the alternative assessment criteria are arranged on a scale with discrete levels ranging from "low" to "very high" (US EPA 2011a).* Following the hazard assessment, DfE evaluates use and exposure data as well, although the focus is on chemical hazards (US EPA 2011, 2014e). The assessment is useful to industry as an informational tool to compare the costs, utility, and risks of chemicals and their potential alternatives. Wal-Mart announced in September 2013 that it will work with suppliers to eliminate ten "priority substances" from its products and replace them with viable alternatives (Grossman 2013). The retailer also plans to begin labeling Wal-Mart brand cleaning products under DfE criteria. DfE alternative assessments have been conducted or are underway for several flame retardants that have raised PBT concerns, such as penta-BDE (US EPA 2014f), deca-BDE (US EPA 2014g), and hexabromocyclododecane (HBCDD) (US EPA 2014h).

As in the labeling requirements, assessments aim to provide industry with information on chemicals' properties (including persistence, bioaccumulation, and toxicity) to facilitate informed substitution decisions based on comparisons of hazard potential. Although the scope of the DfE program is quite limited in the number of chemicals it has addressed through alternatives assessments, it is nonetheless notable for its consistency with the risk management principles. The underlying aim of the alternatives assessment

---

* DfE hazard assessment provides "low," "moderate," or "high" grades for most hazard endpoints.

program is to fulfill the rational alternatives principle by facilitating informed chemical substitution decisions that consider risk–risk tradeoffs. Priorities are determined by stakeholder interest and the potential for an assessment to have a meaningful impact. PBT properties are not a prioritization factor; however, alternatives assessment partnerships exist for several PBTs, including PBDE and HBCDD flame retardants. While the assessment process does not require the generation of new data to fill gaps, the process does facilitate the consideration of the highest quality data available through tiered information gathering, scientific review, and multistakeholder involvement. By considering the functionality of prospective alternatives and potential exposures throughout chemicals' lifecycles, the assessments are also consistent with the exposure and risk assessment and differentiation in use principles.

2. *The PBT Profiler*

Another way that US EPA works with stakeholders to facilitate voluntary risk management is to provide evaluative tools to assist in private actors' decision making (Auer and Alter 2007; Tickner et al. 2013). As part of its Sustainable Future Initiative, US EPA has developed an online PBT-prediction tool for stakeholders called the PBT Profiler (US EPA 2011b). The tool estimates a chemical's persistence, bioaccumulation, and toxicity potential based on its molecular structure. Users provide information on a chemical, and the model compares its PBT potential with the PBT criteria under the TSCA § 5 new chemical program and the TRI final reporting rule (discussed in Chapter 5). The profiler is an example of a binning approach—either a chemical satisfies the criteria to be considered a PBT or it does not (Rorije et al. 2011). The tool does not produce a numerical output but rather provides a color-coded output to show users whether the thresholds for persistence, bioaccumulation, and toxicity are not exceeded (green), exceeded (orange), or greatly exceeded (red) (US EPA 2012b).

The profiler is intended to advance green chemistry by providing PBT information to companies that are developing or considering the use of new chemicals. A company can use the tool to compare the properties of various chemicals it is considering and, in theory, choose the chemical that presents the lowest level of concern, consistent with the rational alternatives principle. The PBT Profiler has reportedly been one of the most widely used chemical screening tools by industry worldwide (Auer and Alter 2007; Chynoweth 2008).

B. *Private programs to encourage voluntary risk management*

1. *Clean Production Action's Green Screen for Safer Chemicals*

The Green Screen for Safer Chemicals is a chemical hazard assessment method that was developed by the NGO Clean Production Action for use by government agencies, businesses, and NGOs (CPA 2014a). Green Screen is based on the principles of green chemistry and the DfE alternatives assessment method. Green Screen has been used in a variety of contexts by regulatory agencies, retailers, product developers, certification programs, and researchers to screen chemicals for hazard endpoints, including PBT properties (Lavoie et al. 2010; CPA 2014b). The method is designed to facilitate the comparison of potential substitutes so that users may select chemicals that present the lowest possible hazard.

The process involves three steps. The first step calls for users to collect and input data for 18 hazard endpoints, including persistence, bioaccumulation, acute and chronic aquatic toxicity, and other common human health and physical hazards (e.g., CMR endpoints) (CPA 2014c). Users may collect data by searching applicable literature for measured data values or values generated by estimation models. These data are then compared with specified hazard criteria, and the program assigns a hazard classification of very low through very high for each endpoint based on weight of evidence judgments (see Rossi and Heine 2007).

Second, users apply specified benchmarks, essentially guidelines, to assist in decision making. Green Screen associates hazard classifications for each endpoint with one of four benchmark scores. Chemicals that receive the highest hazard classifications fall into Benchmark 1. Green Screen recommends that users avoid using these chemicals. The system suggests that users can use Benchmark 2 chemicals, but they should search for less hazardous substitutes as well. Green Screen indicates that Benchmark 3 chemicals should be used, but room for improvement exists. Finally, Benchmark 4 indicates safe, preferred chemicals. PBTs fall into Benchmark 1 or 2 depending on the hazard classification level for each endpoint. For example, a high classification for all three endpoints yields a Benchmark 1 classification, while a moderate classification for all three would yield a Benchmark 2 classification. Notably, the system also identifies some partial PBTs as Benchmark 1, including those that are classified as very high for ecotoxicity and very high for either persistence or bioaccumulation (CPA 2014a).

Third, users apply the information to make more informed decisions on chemicals. The Green Screen tool provides users not only with a benchmark score and its associated advice but also with output tables that summarize hazard classification levels for each endpoint as well as a more detailed report on the known human health and environmental data and data gaps for a chemical. Users can compare the results of potential substitute chemicals to determine which is preferable in terms of hazard properties.

While Green Screen does not constitute a PBT policy per se, it is a good example of a privately developed evaluative tool that government, industry, and NGOs can use to screen and compare chemicals. Although it does not necessarily consider exposure, the Green Screen decision-making framework is not analogous to a regulation that automatically bans PBTs already in commerce without considering exposure. The purpose of the tool is not to determine which chemicals should be managed, but rather to enable users to compare hazards of chemical substitutes such that they can select the less hazardous chemical. Thus, Green Screen and other tools like it facilitate more informed decision making on chemical inputs.

2. *International ChemSec's SIN List*

The International ChemSec is an international NGO whose charter is to advocate against the production and use of hazardous chemicals. ChemSec highlights health and environmental risks of hazardous chemicals and makes the information available to businesses, governments, other NGOs, and the general public. The organization's goals are to influence the Candidate List process under REACH, to provide guidance in chemical substitution decisions to private companies and consumers, and to generally make data on chemicals publicly available (Ligthart 2010).

To that end, ChemSec has used REACH's PBT, vPvB, CMR, and substance of equivalent concern criteria to create a SIN List of chemicals that it believes meet the EU's SVHC criteria and should therefore be phased out under REACH and other regulatory programs. As of October 2014, there are 830 chemicals on the SIN List.* ChemSec developed the list by screening chemicals listed within various registries (e.g., the Canadian DSL categorization results), evaluating existing scientific literature, and using a variety of estimation and modeling packages (e.g., Estimation Program Interface Suite) to determine if the chemicals meet the REACH SVHC criteria (ChemSec 2014a). The methods used to compile the original SIN List have been reviewed and validated by researchers at the Technical University of Denmark (Eriksson, Lützhøft, and Ledin 2009).

Only a small fraction of current SIN List chemicals are PBTs and vPvBs. The list identifies 24 chemicals as PBTs, 4 as vPvBs, and 7 chemicals that fall into both categories. Several of these chemicals are listed because their degradation products raise PBT concerns, congeners are listed separately (e.g., alpha-, beta-, and gamma-congeners of HBCDD), and several are labeled "potential" PBTs and vPvBs. Many of the PBTs and vPvBs that appear on the list are already formally recognized as such by the EU.

The SIN List is an example of a policy approach whereby a private actor provides information to government, industry, NGOs, and the public. The SIN List has been used by scholars as one of many lists of chemicals to examine for various properties, including PBT properties (e.g.,

---

* The SIN List is available online at http://sinlist.chemsec.org.

Howard and Muir 2010; Rorije et al. 2011; Scheringer et al. 2012; Strempel et al. 2012). Businesses have also used the list. ChemSec facilitates a business group—a discussion network through which downstream companies including retailers and products manufacturers can share ideas on chemical substitution (ChemSec 2014b). The major European retail chain Carrefour developed a "pre-list" of suspected SVHCs and distributed it along with the SIN List to its suppliers to encourage the consideration of substitution (*Chemical Watch* 2009).

It is not clear what the precise impact of the SIN List is on regulatory and business decisions. Nonetheless, it seems clear that businesses have noticed it. To the extent that the SVHC determinations are valid, the list could be helpful as one among many guides for business and consumers. However, it is not clear that the list by itself is consistent with the rational alternatives principle or the exposure and risk assessment principle. The list only identifies chemical hazards. It does not provide information on uses or exposure pathways, nor does it identify less hazardous alternatives. Before substitution decisions are made, the risks of using chemical substitutes should be weighed against the risks of using a particular chemical of concern. Otherwise, businesses might be simply trading one risk for another. Thus, the SIN List, and lists like it, only provide one piece of important information and should not constitute the sole basis for decisions on chemical substitution.

What is clear from the DfE, PBT Profiler, GreenScreen, and SIN List experiences is that businesses—products manufacturers and retailers in particular—are searching for tools and sources of information so that they can make decisions about what products they want to sell to consumers. Indeed, retail regulation seems to be on the rise as downstream businesses and consumers incorporate more information on chemicals into their production and purchasing decisions (e.g., Everts 2010; Cutting et al. 2011). In the next section, we provide a case study that demonstrates how government and retail regulation have together influenced the market for flame retardants.

IV.   Case study: State regulation, market forces, retail regulation, and brominated flame retardants

In 1975, California enacted Technical Bulletin 117 (TB 117) establishing flammability standards for the filling materials used in household furnishings, such as couches, pillows, and mattresses. These filling materials included cellular materials, expanded polystyrene beads, shredded and loose fill materials like feathers and down, and synthetic fiber materials. The goal of the flammability standards was to reduce the number of house fires caused by cigarettes igniting furnishings. TB 117 requires furniture fill material to pass a flammability test in which the material is exposed to an open flame for 12 seconds and must not ignite within that time. To meet these standards, furniture manufacturers turned to flame retardant chemicals. Although TB 117 applied only to California, manufacturers applied flame retardants to all household furnishings, regardless of the destination for retail sale. Over the years, flame retardant chemicals were added to other products, such as many infant items and even children's pajamas.

A. *Polybrominated diphenyl ethers*

Some of the more widely used flame retardants belong to a category of chemicals called PBDEs. The number and position of bromine atoms within the molecules can vary, and these variations are called congeners. There are 209 possible congeners of PBDEs. PBDEs are additive flame retardants, meaning the chemical is added to the targeted product but not chemically bound to the product. The flame retardant characteristic of PBDEs results from the process of flame termination during bromine radical formation (de Boer 2004). Commercial formulations contain a mixture of congeners rather than single congeners. The most common formulations in the United States include the following trade names: DE-60F, DE-61, DE-62, and DE-71 for penta-BDE (five bromines) mixtures; DE-79 for octa-BDE (eight bromines) mixtures; and DE 83R and Saytex 102E for deca-BDE (ten bromines) mixtures.

In the late 1990s and early 2000s, evidence of the presence of PBDEs in the environment, biota, and humans was mounting. The most striking evidence was the 50-fold increase in the levels of PBDEs in banked milk in Sweden starting from the early 1970s (Meironyté, Norén, and Bergman 1999). Levels of PBDEs in the North American population were significantly higher than in other regions of the world, suggesting a more widespread use of these chemicals in North America. Hites (2004) confirmed the early findings and showed that PBDE concentrations in humans increased by approximately a factor of 100 over a 30-year period.

Studies confirming the presence of PBDEs in a variety of environmental matrices continued to accumulate, and special attention was given to human exposure. PBDEs were found in human neonatal and maternal blood, breast milk, and adipose tissue samples. In the United States, samples from humans had a median concentration around 35 ng/g lipid, with some samples above 300 ng/g lipid. In Sweden, only neonatal blood and maternal blood were measured, and the approximate concentration for those samples was 2 ng/g lipid. These growing levels indicated increased production and use of PBDEs, and they highlighted concern about the persistent, bioaccumulative, and toxic nature of PBDEs. By the mid-2000s, it was clear that PBDEs had become ubiquitous. They were detected in indoor and outdoor air in the United Kingdom, United States, Canada, and Sweden; in marine mammals such as seals and porpoises in the United Kingdom, Canadian Arctic, Sweden, and the United States; in birds' eggs from Sweden and the United States; in fish from all over the world; and in sediments from Asia, Europe, and North America (Hites 2004).

In 2006, US EPA published a report summarizing animal studies of various PBDE commercial mixtures and individual congeners (US EPA 2006). These studies suggested potential concerns about liver toxicity, thyroid toxicity, developmental toxicity, and developmental neurotoxicity. PBDEs were also added to the growing list of chemicals linked to obesity and metabolic disorders (Lim, Lee, and Jacobs 2008).

B.   Government regulation and voluntary phase-out

By 2004, the evidence of PBDEs being persistent and bioaccumulative, especially penta-BDE and octa-BDE, had mounted, and PBDEs were getting significant attention outside the scientific literature. Great Lakes Chemical Corporation (now Chemtura Corporation) was the only manufacturer of PBDEs in the United States. In response to public concern, and the likelihood of government regulation, Great Lakes Chemical Corporation voluntarily ceased manufacturing of the penta-BDE and octa-BDE commercial mixtures. The voluntary phase-out did not affect the importation of goods containing these compounds, nor did it affect production of deca-BDE, which remained on the market until the end of 2013.

Eventually, US EPA addressed PBDEs through a significant new use rule, or SNUR (Duvall 2014). The SNUR required anyone who intended to manufacture or import a chemical or mixture containing any of the congeners present in commercial mixtures of penta-BDE or octa-BDE to notify US EPA at least 90 days in advance. The SNUR did not address importation of articles to which penta-BDE or octa-BDE had been added. To complement the voluntary phase-out and the SNUR, the DfE program provided data to inform substitution to safer alternatives (Auer and Alter 2007). Parallel to US EPA's SNUR regulations, some states regulated PBDEs independently. In 2003, California passed a ban on penta-BDE and octa-BDE, effective in 2008, which eliminated "manufacturing, processing, distributing a product, or a flame-retarded part of a product, containing more than one-tenth of 1 percent" of either congener (Daub 2005).* As noted previously, the state of Oregon banned the introduction of penta-BDE and octa-BDE into commerce in 2005. At least nine other US states have implemented PBDE bans or regulations, as have many countries in

* California Assembly Bill 302, Section 1, Chapter 10, Section 108922(a).

the EU. In 2009, commercial penta-BDE was added to Annex A (elimination) of the Stockholm Convention on Persistent Organic Pollutants.

Despite these measures, PBDEs continued to be widely detected in humans and the environment (US EPA 2009; Duvall 2014). In addition to the existing stock of goods containing PBDEs already present in the United States (the shelf life of a sofa could be as long as 10 to 20 years), new products containing PBDEs have continued to be imported. PBDEs are extremely persistent in the environment, and evidence suggests that some congeners such as deca-BDE can break down into lower-level congeners, which are often more persistent than the parent compound (US EPA 2009).

C.   Alternatives to PBDEs

After the voluntary withdrawal of PBDEs, companies needing to meet the flammability standards of TB117 replaced PBDEs with alternative chemicals that, in general, have structures and properties very similar to PBDEs. The majority of these chemicals were registered under the TSCA Inventory, and manufacturers were therefore not required to provide any new testing data before introducing them into the market.

In 2004, to replace the commercial penta-BDE mixture, the flame retardant industry began to use alternative formulations called Firemaster 550, Firemaster BZ-54, and DP-45. Firemaster 550 consists of about 35% of 2-ethylhexyl-2,3,4,5-tetrabromobenzoate (TBB), about 15% of bis(2-ethylhexyl)-tetrabromophthalate (TBPH), and about 50% of aromaticphosphate esters. Firemaster BZ-54 consists of about 70% of TBB and about 30% of TBPH. DP-45 contains TBPH only (Bearr, Stapleton, and Mitchelmore 2010; Ma, Venier, and Hites 2012). TBB and TBPH share some of the unwanted properties of the compounds they are replacing. For example, they can accumulate in fish and cause DNA damage (Bearr, Stapleton, and Mitchelmore 2010). TBB and TBPH were first detected in 2011 in polyurethane foam collected from baby products (Stapleton et al. 2011). Shortly after, an analysis of air samples collected on the shores of the Great Lakes showed not only that these chemicals were present at levels similar to those of penta-BDEs but also that these levels were doubling about every two years (Ma, Venier, and Hites 2012). These findings strongly suggest that TBB and TBPH are currently used as the major replacement of penta-BDEs. From a toxicological standpoint, Firemaster 550 was recently found to be an endocrine disruptor and an obesogen at environmentally relevant levels (Patisaul et al. 2013).

Organophosphate flame retardants (OPFRs) are also used as alternatives to PBDEs. While OPFRs in general are considered safer than PBDEs because of their lower persistence and bioaccumulation, those containing chlorine atoms are carcinogenic (for example, tris(2-chloroethyl) phosphate, TCEP). TCEP is the chlorinated analog of a compound that came to the public attention in the late 1970s known simply as Tris (Tris (2,3-dibromopropyl) phosphate). Tris was added to children's pajamas in quantities up to 10% of the product weight. After it was detected in the urine of babies wearing those pajamas and it was found to be a potential carcinogen, it was removed from pajamas. Tris was replaced by TCEP, which is still being used. Stapleton et al. (2012) reported finding several OPFRs in polyurethane foam from several baby products. This market shift in the use of flame retardants is well depicted by Stapleton et al. (2012) in which the authors examined differences in foam samples from sofas manufactured before and after 2004. Samples from sofas produced before 2004 contained PBDEs and only small amounts of OPFRs; foam from sofas purchased after 2004 contained mainly OPFRs and the components of Firemaster 550.

D.   Public concern and retail regulation

As consumers have become more informed about the potential dangers of PBDEs, they have questioned the necessity of such chemicals in their household products. Environmental organizations and grass roots activists have pushed for elimination of PBDEs from consumer products. For example, Safer Chemicals, Healthy Families, an organization representing more than 11 million people and 450 businesses and organizations, is encouraging business to phase out toxic chemicals such as PBDEs. In 2012 and 2013, this group organized a march in Washington, DC, dubbed the National Stroller Brigade, to bring attention to their mission. During the second half of 2012, the *Chicago Tribune* published a six-part investigative series on flame retardants (e.g., Callahan and Roe 2012). This series received considerable attention and won a 2013 Pulitzer Prize in Investigative Reporting. This media attention allowed the public to easily and quickly learn about the potential risks of brominated flame retardants and the occurrence of these chemicals in consumer products.

Even before some of these media events, US retailer Wal-Mart began to eliminate PBDEs from its products (Cutting et al. 2011). In 2011, Wal-Mart notified suppliers that as of June 1 of that year, the company would begin testing products to ensure compliance with their no-PBDE stance. According to a *Washington Post* article, Wal-Mart was prompted to take these measures as US states began to ban PBDEs (Layton 2011). Other retail firms and manufacturers have taken similar steps. Dell, a major manufacturer of computers and electronics, removed PBDEs from all of its products in 2002, complying with the EU's Restriction of Hazardous Substances directive four years before its enactment. In 2012, Dorel Juvenile Group, a large manufacturer of children's products in the United States, announced that it had established internal policies prohibiting the use of PBDEs (DJG 2012). In 2014, Kaiser Permanente, an integrated managed care consortium based in California, announced that it will stop purchasing furniture treated with flame retardants (Kaiser Permanente 2014). According to its website, Kaiser Permanente buys approximately $30 million of furniture each year. Lastly, the Business and Institutional Furniture Manufacturers Association released a position paper in 2012 strongly supporting the elimination of flame retardants from all of its manufactured products (BIFMA 2012).

Voluntary initiatives by manufacturers and retailers in the United States have occurred in advance of any federal regulations, although they are supported by DfE alternatives assessment partnerships on PBDEs. As mentioned above, this phenomenon has been called "retail regulation" and is viewed by some as preferable to the lengthy process of developing and implementing federal regulations for PBDEs (e.g., Layton 2011; Grossman 2013; Bomgardner 2014). As a result of pressure from the public and from several environmental groups, California revised TB117 for better fire safety without the need for flame retardant chemicals. On January 1, 2015, TB117-2013 went into effect and established a new flammability test for the fill material used in furniture.* Furniture makers can meet the requirement without using flame retardants, although TB117-2013 does not itself preclude the use of PBDEs or other chemical flame retardants.

It appears that a retail regulation approach bolstered by state-level regulations and the DfE alternatives assessment program has succeeded in eliminating PBDEs from many consumer products, but it is unclear if this approach would be successful for other PBTs. The case of PBDEs likely captured public interest because human exposure was through ubiquitous and seemingly harmless consumer products, such as sofas, pillows, and other furniture. Studies of PBDE accumulation in humans, and in particular human breast milk, were available and highlighted by the media. These types of nonstandard data (with respect to PBT determinations) are unlikely to be available for most potential PBTs. Furthermore, most potential PBTs are unlikely to receive the intensive study by academic scientists that PBDEs received because the chemicals have lower production volumes or are not used in household furnishings or children's products. So while advocacy groups might view retail regulation as a promising means for elimination of a particular chemical, we expect that most potential PBTs will lack appropriate data and exposure pathways to capture the public attention necessary for retail regulation to be effective.

---

* 4 California Code of Regulations §§ 1373.2, 1374.

V.   Conclusion

The case of PBDEs illustrates how risk management decision making by private actors operates in the context of subnational and national risk management frameworks. Often, those frameworks are explicitly designed to encourage voluntary risk management, such as US EPA's DfE program. Voluntary risk management might also be undertaken in anticipation of regulatory action or to convince authorities that regulatory action is not necessary. Private retail regulation decisions are encouraged by a host of other factors as well, including concerns raised by highly salient scientific research (e.g., Hites 2004) and investigative reporting (e.g., Callahan and Roe 2012).

In addition to programs designed to encourage voluntary risk management, there is a substantial amount of regulatory activity on chemicals occurring at the state level in the United States, especially targeting PBTs. Most of these programs tend to be limited in scope given the heavy resource constraints on environmental policy implementation at the subnational level. However, as subnational programs expand in number and variety, the transaction costs associated with risk management for firms that do businesses in national and international marketplaces might become onerous. At the very least, subnational governments that are enacting regulations pertaining to PBTs should take care to harmonize their approaches in a coordinated effort. In the final chapter that follows, we describe in detail our findings and recommendations on PBT science and policy, including a call for greater harmonization in approaches to PBT determinations and policies.

References

Allen, J.H. 2013. "The Wicked Problem of Chemicals Policy: Opportunities for Innovation." *Journal of Environmental Studies and Science* 3:101–8.

Auer, C., and J. Alter. 2007. "The Management of Industrial Chemicals in the USA." In *Risk Assessment of Chemicals: An Introduction*, edited by C.J. van Leeuwen, and T.G. Vermeire, 553–74. Dordrecht, The Netherlands: Springer.

Bearr, J.S., H.M. Stapleton, and C.L. Mitchelmore. 2010. "Accumulation and DNA Damage in Fathead Minnows (*Pimephales promelas*) Exposed to 2 Brominated Flame-Retardant Mixtures, Firemaster® 550 and Firemaster® BZ-54." *Environmental Toxicology and Chemistry* 29:722–9.

BIFMA (Business + Institutional Furniture Manufacturers Association). 2012. "Elimination of Fire Retardant Chemicals in Office Furniture Products," March 27. Available at https://c.ymcdn.com/sites/www.bifma.org/resource/resmgr/advocacy/position_paper_fr_march_27_1.pdf.

Bomgardner, M.M. 2014. "Walmart and Target Take Aim at Hazardous Ingredients." *Chemical & Engineering News* 92:19–21.

Callahan, P., and S. Roe. 2012. "Fear Fans Flames for Chemical Makers." *Chicago Tribune*, May 6.

Castrilli, J.F., and F. de Leon. 2014. "Ontario's Living List—A Dead Thing?" *Canadian Environmental Law Association Blog*, May 20. Available at http://www.cela.ca/blog/2014-05-20/ontarios-living-list-dead-thing.

CCSCE (Center for Continuing Study of the California Economy). 2014. "California Once Again the World's 8th Largest Economy," July. Available at http://www.ccsce.com/PDF/Numbers-July-2014-CA-Economy-Rankings-2013.pdf.

Chemical Watch. 2009. "Retailers Integrate REACH into Chemicals Management." Global Business Briefing, February.

ChemSec. 2014a. "Comprehensive Methodology for QON the SIN List: From 2008 until 2014," October. Available at http://www.chemsec.org/images/stories/2014/Full_SIN_Methodology_October_2014.pdf.

ChemSec. 2014b. "A Tool for Businesses." Accessed December 29, 2014. Available at http://www.chemsec.org/what-we-do/sin-list/about-sin/users/businesses.

Chynoweth, E. 2008. "Downstream Users Take Up EPA Sustainable Futures Tools." *Chemical Watch.* Global Business Briefing, September.

CIEL (Center for International Environmental Law). 2005. *U.S. States and the Global POPs Treaty: Parallel Progress in the Fight against Toxic Pollution.* Washington, DC: CIEL. Available at http://www.ciel.org/Publications/States_POPs_May05.pdf.

Cowan, D.M., T. Kingsbury, A.L. Perez, T.A. Woods, M. Kovochich, D.S. Hill, A.K. Madl, and D.J. Paustenbach. 2014. "Evaluation of the California Safer Consumer Products Regulation and the Impact on Consumers and Product Manufacturers." *Regulatory Toxicology and Pharmacology* 68:23–40.

CPA (Clean Production Action). 2014a. "GreenScreen for Safer Chemicals," January. Available at http://www.greenscreenchemicals.org/static/ee_images/uploads/resources/2pager_greenscreen_2014.pdf.

CPA (Clean Production Action). 2014b. "How is GreenScreen Used?" Accessed December 29. Available at http://www.greenscreenchemicals.org/practice/how-is-gs-used.

CPA (Clean Production Action). 2014c. "Full GreenScreen Method." Accessed December 29. Available at http://www.greenscreenchemicals.org/method/full-greenscreen-method.

Cutting, R.H., L.B. Cahoon, J.F. Flood, L. Horton, and M. Schramm. 2011. "Spill the Beans: GoodGuide, Walmart and EPA Use Information as Efficient, Market-Based Environmental Regulation." *Tulane Environmental Law Journal* 24:291–334.

Daub, T. 2005. "California—Rogue State or National Leader in Environmental Regulation? An Analysis of California's Ban of Brominated Flame Retardants." Note. *Southern California Interdisciplinary Law Journal* 14:345–70.

de Boer, J. 2004. "Brominated Flame Retardants in the Environment—The Price for Our Convenience?" *Environmental Chemistry* 1:81–5.

DEQ (Department of Environmental Quality, State of Oregon). 2009. "Senate Bill 7373: Development of a Priority Persistent Pollutant (P³) List for Oregon." Executive Summary. Available at http://www.deq.state.or.us/wq/SB737/docs/P3LrepExecutiveSum.pdf.

DEQ (Department of Environmental Quality, State of Oregon). 2011. "Fact Sheet: Implementing Senate Bill 737: June 2011 Update." Available at http://www.deq.state.or.us/wq/pubs/factsheets/programinfo/09-WQ-25SB737FS.pdf.

DEQ (Department of Environmental Quality, State of Oregon). 2012. *DEQ Toxics Reduction Strategy: Descriptions of Actions.* Salem, ON: Kevin Masterson, November. Available at http://www.deq.state.or.us/toxics/docs/Toxics StrategyNov28.pdf.

DEQ (Department of Environmental Quality, State of Oregon). 2014. "Reducing Toxics in Oregon." Accessed December 29, 2014. Available at http://www .oregon.gov/deq/Pages/ToxicsReduction.aspx.

Ditz, D.W. 2007. "The States and the World, Twin Levels for Reform of U.S. Federal Law on Toxic Chemicals." *Sustainable Development Law & Policy* 8:27–30.

DJG (Dorell Juvenile Group). 2012. "Dorel Juvenile Group Prohibits Potentially Toxic Flame Retardants in Children's Products," July 26. Available at http:// www.djgusa.com/usa/eng/News/Detail/534-Dorel-Juvenile-Group -Prohibits-Potentially-Toxic-Flame-Retardants-in-Childrens-Products.

DOE (Department of Ecology, Washington State). 2000. *Proposed Strategy to Continually Reduce Persistent, Bioaccumulative Toxics (PBTs) in Washington State.* Available at https://fortress.wa.gov/ecy/publications/publications/0003054.pdf.

DOE (Department of Ecology, Washington State). 2007. *Multiyear PBT Chemical Action Plan Schedule.* Publication No. 07-07-016. Available at https://fortress .wa.gov/ecy/publications/publications/0707016.pdf.

DOE (Department of Ecology, Washington State). 2014a. "What is a Chemical Action Plan?" Accessed December 28. Available at http://www.ecy.wa.gov /programs/swfa/pbt/caps.html.

DOE (Department of Ecology, Washington State). 2014b. *Draft PCB Chemical Action Plan.* Publication no. 14-07-024. Available at https://fortress.wa.gov/ecy /publications/publications/1407024.pdf.

DOE (Department of Ecology, Washington State) and DOH (Department of Health, Washington State). 2003. *Washington State Mercury Chemical Action Plan.* Available at https://fortress.wa.gov/ecy/publications/publications /0303001.pdf.

DTSC (California Department of Toxic Substances Control). 2013. *Safer Consumer Products: Summary of Revised Proposed Regulations.* R-2011-02, April. Available at https://www.dtsc.ca.gov/LawsRegsPolicies/Regs/upload/5-SCP-Regs _Summary-of-Changes-April-20131.pdf.

DTSC (California Department of Toxic Substances Control). 2014a. *DTSC'S Initial Proposed Priority Products List for the Safer Consumer Products Program,* March 2014. Available at https://www.dtsc.ca.gov/SCP/upload/Proposed-Initial -Priority-Product-List.pdf.

DTSC (California Department of Toxic Substances Control). 2014b. *Safer Consumer Products Draft Priority Product Work Plan: Three Year Work Plan,* September. Available at https://www.dtsc.ca.gov/SCP/upload/FINAL-DRAFT-PPWP -140909.pdf.

DTSC (California Department of Toxic Substances Control). 2015. "Candidate Chemical List." Accessed January 2, 2015. Available at https://www.dtsc.ca .gov/SCP/ChemList.cfm.

Duvall, M.N. 2014. "Flame Retardants Face Increasing Federal and State Scrutiny." *National Law Review,* July 25. Available at http://www.natlawreview.com /article/flame-retardants-face-increasing-federal-and-state-scrutiny.

Easthope, T., and L. Valeriano. 2007. "Phase Out of Persistent, Bioaccumulative or Highly Toxic Chemicals." *New Solutions* 17:193–207.

Eriksson, E., H.H. Lützhøft, and A. Ledin. 2009. "Second Opinion on the Hazards Associated with the Substances Selected for the REACH SIN* List 1.0." Department of Environmental Engineering, Miljoevej, B113. Lyngby, Denmark: Technical University of Denmark. Available at http://www.chemsec.org/images/stories/news_publications/SIN_List_DTU_Evaluation_2nd_opinion_Report_final.pdf.

Everts, S. 2010. "Greener Chemistry: Everyday Products with an Eco-Tinge." *New Scientist*. 205:34–8.

Gamper-Rabindran, S., and S.R. Finger. 2013. "Does Industry Self-Regulation Reduce Pollution? Responsible Care in the Chemical Industry." *Journal of Regulatory Economics* 43:1–30.

Greer, L. 2010. Statement before the Subcommittee on Commerce, Trade, and Consumer Production, House Committee on Energy and Commerce, US Congress. *The Toxic Substances Control Act and Persistent, Bioaccumulative, and Toxic Chemicals: Examining Domestic and International Actions: 111th Cong., 2nd sess.*, March 4.

Grossman, E. 2013. "Walmart Targets Ten Substances of Concern in Consumer Products." *Chemical Watch*, September 13.

Hites, R.A. 2004. "Polybrominated Diphenyl Ethers in the Environment and in People: A Meta-Analysis of Concentrations." *Environmental Science & Technology* 38:945–56.

Hogue, C. 2013. "State Lawmakers Introducing Bill to Restrict Chemicals." *Chemical & Engineering News* 91:37–9.

Hogue, C. 2014. "Congress Considers States' Regulations." *Chemical & Engineering News* 92:22.

Hope, B.K., D. Stone, T. Fuji, R.W. Gensemer, and J. Jenkins. 2010. "Meeting the Challenge of Identifying Persistent Pollutants at the State Level." *Integrated Environmental Assessment and Management* 6:735–48.

Howard, P.H., and D.C.G. Muir. 2010. "Identifying New Persistent and Bioaccumulative Organics among Chemicals in Commerce." *Environmental Science & Technology* 44:2277–85.

IC2 (Interstate Chemicals Clearinghouse). 2015. "U.S. State Chemicals Policy Database." Accessed January 2, 2015. Available at http://theic2.org/chemical-policy.

Kaiser Permanente. 2014. "Kaiser Permanente Commits to Purchasing Furniture Free from Toxic Flame Retardant Chemicals." *Press Release*, June 3. Available at http://share.kaiserpermanente.org/article/kaiser-permanente-commits-to-purchasing-furniture-free-from-toxic-flame-retardant-chemicals/#sthash.hT5VtCJY.dpuf.

Keller and Heckman LLP. 2014. "Chemical Legislation in Oregon (Existing and Proposed)." Accessed December 29. Available at https://www.khlaw.com/files/11966_Oregon.pdf.

Lavoie, E.T., L.G. Heine, H. Holder, M.S. Rossi, R.E. Lee II, E.A. Connor, M.A. Vrabel, D.M. DiFiore, and C.L. Davies. 2010. "Chemical Alternatives Assessment: Enabling Substitution to Safer Chemicals." *Environmental Science & Technology* 44:9244–9.

Layton, L. 2011. "Wal-Mart Bypasses Federal Regulators to Ban Controversial Flame Retardant." *Washington Post*, February 26.

LCSP (Lowell Center for Sustainable Production). 2015. "Chemicals Policy & Science Initiative." Accessed January 2. Available at http://www.chemicalspolicy.org/home.php.

Ligthart, J.J. 2010. "The SIN List as Model for the Identification of Substances of Very High Concern." *Journal of Epidemiology and Community Health* 64:654–5.

Lim, J.S., D.H. Lee, and D.R. Jacobs, Jr. 2008. "Association of Brominated Flame Retardants with Diabetes and Metabolic Syndrome in the U.S. Population, 2003–2004." *Diabetes Care* 31:1802–7.

Ma, Y., M. Venier, and R.A. Hites. 2012. "2-Ethylhexyl Tetrabromobenzoate and Bis(2-ethylhexyl) Tetrabromophthalate Flame Retardants in the Great Lakes Atmosphere." *Environmental Science & Technology* 46:204–8.

Margulies, J.B., and W.L. Troutman. 2013. "The Safer Consumer Products Regulations: California's Green Chemistry Initiative." Los Angeles, CA: Fulbright and Jaworsky, LLP. Available at http://www.fulbright.com /e_templates/Data/!Tik/02212013EnvironmentalWP/WhitePaper2.pdf.

Meironyté, D., K. Norén, and Å. Bergman. 1999. "Analysis of Polybrominated Diphenyl Ethers in Swedish Human Milk: A Time-Related Trend Study, 1972–1997." *Journal of Toxicology and Environmental Health, Part A* 58:329–41.

Mitchell, R.R., C.L. Summer, S.A. Blonde, D.M. Bush, G.K. Hurlburt, E.M. Snyder, and J.P. Giesy. 2002. "SCRAM: A Scoring and Ranking System for Persistent, Bioaccumulative, and Toxic Substances for the North American Great Lakes—Resulting Chemical Scores and Rankings." *Human and Ecological Risk Assessment* 8:537–57.

MOE (Ontario Ministry of the Environment). 2008. *Discussion Paper: Creating Ontario's Toxics Reduction Strategy*, August. Available at http://www.down loads.ene.gov.on.ca/envision/env_reg/er/documents/2008/010-4374.pdf.

MOE (Ontario Ministry of the Environment). 2009. "Backgrounder: Development of Lists of Substances Proposed to be Prescribed under the Toxics Reduction Act, 2009: Toxic Substances and Substances of Concern," September 21. Available at https://ia601205.us.archive.org/32/items/stdprod080013.ome /stdprod080013.pdf.

MOE (Ontario Ministry of the Environment). 2012. *Ontario Toxics Reduction Program: A Guide for Regulated Facilities*. Available at https://www .ontario.ca/environment-and-energy/ontario-toxics-reduction-program -guide-regulated-facilities.

MOE (Ontario Ministry of the Environment). 2014. *The Draft Living List Framework under Ontario's Toxics Reduction Program*. Available at http://www.ecolog .com/daily_images/1002963655-1002976367.pdf.

Morgenstern, R., and W.A. Pizer. 2007. *Reality check: The Nature and Performance of Voluntary Environmental Programs in the United States, Europe, and Japan*. Washington DC: RFF Press.

Muir, D.C.G., and P.H. Howard. 2006. "Are There Other Persistent Organic Pollutants? A Challenge for Environmental Chemists." *Environmental Science & Technology* 40:7157–66.

Parametrix. 2008. *SB 737 Background Information Summary of Listing Processes for Persistent, Bioaccumulative, and Toxic Chemicals: Final Report*. Prepared for Oregon Association of Clean Water Agencies and League of Oregon Cities, September. Available at http://www.deq.state.or.us/wq/sb737/docs/PPSWG /FinalPBTReport.pdf.

Patisaul, H.B., S.C. Roberts, N. Mabrey, K.A. McCaffrey, R.B. Gear, J. Braun, S.M. Belcher, and H.M. Stapleton. 2013. "Accumulation and Endocrine Disrupting Effects of the Flame Retardant Mixture Firemaster® 550 in Rats: An Exploratory Assessment." *Journal of Biochemical and Molecular Toxicology* 27:124–36.

Rorije, E., E.M.J. Verbruggen, A. Hollander, T.P. Traas, and M.P.M. Janssen. 2011. "Identifying Potential POP and PBT Substances: Development of a New Persistence/Bioaccumulation-Score." RIVM Report 601356001/2011. Utrecht, The Netherlands: National Institute for Public Health and the Environment. Available at http://www.rivm.nl/bibliotheek/rapporten/601356001.pdf.

Rossi, M., and L. Heine. 2007. *The Green Screen for Safer Chemicals: Evaluating Flame Retardants for TV Enclosures. Clean Production Action.* Somerville, MA: Clean Production Action. Available at http://www.greenscreenchemicals.org/static/ee_images/uploads/resources/EvaluatingFlameRetardants_GreenScreenSaferChemicals_2007.pdf.

Safer States. 2015. "Adopted Policy." Accessed January 2. Available at http://www.saferstates.com/bill-tracker.

Scheringer, M., S. Strempel, S. Hukari, C.A. Ng, M. Blepp, and K. Hungerbuhler. 2012. "How Many Persistent Organic Pollutants Should We Expect?" *Atmospheric Pollution Research* 3:383–91.

Schifano, J.N., J. Tickner, and Y. Torrie. 2009. *State Leadership in Formulating and Reforming Chemicals Policy: Actions Taken and Lessons Learned. Lowell Center for Sustainable Production.* Lowell, MA: University of Massachusetts Lowell. Available at http://www.chemicalspolicy.org/downloads/StateLeadership.pdf.

Snyder, E.M., S.A. Snyder, J.P. Giesy, S.A. Blonde, G.K. Hurlburt, C.L. Summer, R.R. Mitchell, and D.M. Bush. 2000. "SCRAM: A Scoring and Ranking System for Persistent, Bioaccumulative, and Toxic Substances for the North American Great Lakes." *Environmental Science and Pollution Research* 7:176–84.

Stapleton, H.M., S. Klosterhaus, A. Keller, P.L. Ferguson, S. van Bergen, E. Cooper, T.F. Webster, and A. Blum. 2011. "Identification of Flame Retardants in Polyurethane Foam Collected from Baby Products." *Environmental Science & Technology* 45:5323–31.

Stapleton, H.M., S. Sharma, G. Getzinger, P.L. Ferguson, M. Gabriel, T.F. Webster, and A. Blum. 2012. "Novel and High Volume Use Flame Retardants in US Couches Reflective of the 2005 PentaBDE Phase Out." *Environmental Science & Technology* 46:13432–9.

Strempel, S., M. Scheringer, C.A. Ng, and K. Hungerbühler. 2012. "Screening for PBT Chemicals among the 'Existing' and 'New' Chemicals of the EU." *Environmental Science & Technology* 46:5680–7.

Tickner, J.A., K. Geiser, C. Rudisill, and J.N. Schifano. 2013. "Alternatives Assessment in Regulatory Policy: History and Future Directions." In *Chemical Alternatives Assessments*, edited by R.E. Hester, and R.M. Harrison, 256–95. Cambridge, UK: Royal Society of Chemistry.

US EPA (United States Environmental Protection Agency). 2006. *Polybrominated Diphenyl Ethers (PBDEs) Project Plan*, March. Available at http://www.epa.gov/oppt/existingchemicals/pubs/actionplans/proj-plan32906a.pdf.

US EPA (United States Environmental Protection Agency). 2009. *Polybrominated Diphenyl Ethers (PBDEs) Action Plan*, December 30. Available at http://www.epa.gov/oppt/existingchemicals/pubs/actionplans/pbdes_ap_2009_1230_final.pdf.

US EPA (United States Environmental Protection Agency). 2011a. *Design for the Environment Program Alternatives Assessment Criteria for Hazard Evaluation: Version 2.0.* Office of Pollution Prevention and Toxics, August. Available at http://www.epa.gov/dfe/alternatives_assessment_criteria_for_hazard_eval.pdf.

US EPA (United States Environmental Protection Agency). 2011b. "PBT Profiler." Last modified April 18. Available at http://www.epa.gov/pbt/tools/tool box.htm.

US EPA (United States Environmental Protection Agency). 2012a. *Design for the Environment Program Master Criteria for Safer Ingredients: Version 2.1.* Office of Pollution Prevention and Toxics, September. Available at http://www.epa .gov/dfe/pubs/projects/gfcp/dfe_master_criteria_safer_ingredients_v2_1 .pdf.

US EPA (United States Environmental Protection Agency). 2012b. "Estimating Persistence, Bioaccumulation, and Toxicity Using the PBT Profiler." In *Sustainable Futures/P2 Framework Manual 2012 EPA-748-B12-001.* Washington, DC: Office of Pollution Prevention and Toxics, EPA. Available at http:// www.epa.gov/oppt/sf/pubs/pbt-profiler.pdf.

US EPA (United States Environmental Protection Agency). 2014a. "Design for the Environment (DfE): About Us." Last modified December 22. Available at http://www.epa.gov/dfe/pubs/about/index.htm.

US EPA (United States Environmental Protection Agency). 2014b. "Design for the Environment (DfE): Best Practices." Last modified December 5. Available at http://www.epa.gov/dfe/best_practices.html.

US EPA (United States Environmental Protection Agency). 2014c. "What does the DfE Label Mean?" Last modified December 5. Available at http://www.epa .gov/dfe/pubs/projects/formulat/label.htm.

US EPA (United States Environmental Protection Agency). 2014d. "DfE's Standard and Criteria for Safer Chemical Ingredients." Last modified December 5. Available at http://www.epa.gov/dfe/pubs/projects/gfcp/#Toxicity.

US EPA (United States Environmental Protection Agency). 2014e. "Alternatives Assessments." Last modified December 5. Available at http://www.epa .gov/dfe/alternative_assessments.html.

US EPA (United States Environmental Protection Agency). 2014f. "Environmental Profiles of Chemical Flame-Retardant Alternatives for Low-Density Polyurethane Foam." Last modified December 5. Available at http://www .epa.gov/dfe/pubs/flameret/ffr-alt.htm.

US EPA (United States Environmental Protection Agency). 2014g. "Flame-Retardant Alternatives for DecaBDE Partnership." Last modified December 5. Available at http://www.epa.gov/dfe/pubs/projects/decaBDE/.

US EPA (United States Environmental Protection Agency). 2014h. *Flame Retardant Alternatives for Hexabromocyclododecane HBCD.* Final Report. EPA Publication 740R14001, June. Available at http://www.epa.gov/dfe/pubs/projects/hbcd /hbcd-full-report-508.pdf.

chapter seven

PBT determinations and policies
Findings and recommendations

To guide our findings and recommendations, we employed seven management principles that have proven to be useful on a wide range of technologies. Discussed in Chapter 1, those principles address information quality, precaution, priority setting, exposure and risk assessment, differentiation of applications, rational alternatives, and the value of information. For each finding and recommendation, we refer to the specific management principles that are relevant to the finding or recommendation.

I.   Findings and recommendations

A.   Findings on PBT determinations

1. While it was originally thought that the number of PBTs in commerce was fairly small (perhaps fewer than 20), recent screening exercises of chemical inventories suggest that the number of PBTs in commerce is probably between 100 and 1000, depending on the particular data and cutoff values that are used. This number is small relative to the total number of chemicals in commerce but large enough to put substantial work burdens on regulatory agencies and industry.

   The PBT concept arose mainly from the concerns over the "dirty dozen" chemicals. The PBT characteristics were recognized and grouped together to identify similar chemicals, which at the time were thought to be at most a few dozen. However, studies have shown the number of potential PBTs to be much larger, most likely between 100 and 1000 depending upon the criteria and screening method.

   As policymakers, nongovernmental organizations (NGOs), and industry consider giving more regulatory priority to PBTs, it is important to consider the resource implications for regulatory agencies, industry, and the NGO community, since all three of these sectors will be expected to devote more of their scarce resources to PBTs. If the number of PBTs in commerce proves to be 1000 or greater, then it may be necessary to develop priority-setting schemes for PBTs, to ensure that those posing the greatest potential risk are evaluated first.

Relevant principles: information quality, precaution, priority setting, value of information

2. The estimated number of PBTs in commercial use is sensitive to the cutoff values employed and the specific tests and models used for PBT determination. When modeling data are replaced by appropriate measurements, the estimated number of PBTs tends to decline significantly.

There are discrepancies in the persistence, bioaccumulation, and toxicity cutoff values for PBT determinations and, as a result, the number of PBTs will vary across organizations that are using different PBT determination procedures. Thus, if one jurisdiction reports a large number of PBTs and another jurisdiction reports a small number of PBTs, the explanation may simply be that the two jurisdictions are using different cutoff values for persistence, bioaccumulation, and/or toxicity.

Moreover, organizations that produce lists of PBTs do not necessarily assess the same starting group of chemicals. If organization A starts with a larger inventory of chemicals (e.g., with fewer exclusions) than organization B does, then we should not be surprised if organization A produces a longer list of potential PBTs. For example, variation arises from the inclusion or exclusion of heavy metals and individual congeners of large classes of chemicals, such as PCBs.

When a chemical appears on one list of PBTs but not on another list, there are a variety of possible explanations. A chemical might not appear on a particular list of PBTs simply because that organization did not include the chemical in its starting universe of substances, or the organization may not have assessed it yet and made a determination. Thus, the fact that a chemical is found on one list of PBTs but not on another cannot, without more inquiry, suggest anything particularly important.

Finally, when the results of modeling are replaced by measured values, many potential PBTs are not listed as PBTs. The experience in both Europe and Canada is consistent in this regard. Insofar as the modeling exercises are intended to be conservative (i.e., they may err on the side of minimizing false-negatives in PBT designations), we should not be surprised when measured values produce fewer PBTs than modeling exercises.

Relevant principles: information quality, precaution, priority setting, differentiation

3. Instead of making PBT determinations based on numeric criteria, the trend is to make them based on weight of evidence (WOE), as is common practice in toxicity assessment. While this trend is preferable because it allows all relevant scientific data to be considered, little guidance exists on how WOE judgments about persistence and bioaccumulation should be made on a case-by-case basis. As a result, PBT determinations may become, to varying degrees, subjective, unpredictable, inconsistent, and not fully replicable.

With the revision of Registration, Evaluation, Authorization, and Restriction of Chemicals (REACH) Annex XIII, WOE terminology has come to the forefront of PBT determination. Strict cutoff values are useful, but there are a number of chemicals for which blind adherence to cutoff values will produce dubious results. Thus, the PBT classification process is not straightforward in all cases. To address these issues, WOE is used to determine PBTs on a case-by-case basis. In the long run, WOE should provide more accurate determinations, but there is currently a lack of transparency about these determinations. WOE guidelines are common in risk assessment (e.g., in carcinogenicity determinations), but very little guidance is specific to PBTs. There are specific challenges with PBT determinations based on WOE because of the combination of three different characteristics (persistence, bioaccumulation, and toxicity). When WOE is applied without guidance, concerns can arise about subjectivity, credibility, and predictability.

Relevant principles: information quality, value of information

4. PBT determinations are hybrid decisions based partly on scientific knowledge and partly on science-policy judgments (e.g., the selection of cutoff values from continuous distributions of persistence, bioaccumulation, and toxicity).

There is a strong scientific rationale for concern about the three individual characteristics of persistence, bioaccumulation, and toxicity. The decision to combine the three properties into a new chemical classification—the official PBT determination—was a policy decision made in some countries and programs but not in others.

When cutoff values are chosen for persistence, bioaccumulation, and toxicity, they appear to be based, at least in part, on policy decisions aimed at capturing a desired number of chemicals. Under the Stockholm Convention, for example, the cutoff values seem to have been selected to capture a small number of chemicals with PBT properties similar to the original dirty dozen. Under REACH, the cutoff values were selected to capture a larger number of chemicals, reflecting a more expansive policy ambition. There is a strong scientific rationale for examining the persistence, bioaccumulation, and toxicity characteristics of a substance, but decision makers should recognize that the selection of specific cutoff values is primarily a policy decision. As a result, the distinction between PBTs and non-PBTs reflects a hybrid science-policy decision-making process rather than a precise scientific finding.

Relevant principles: information quality, precaution, priority setting

5. Some governmental organizations have produced lists of PBTs, while other organizations simply use information about the properties of persistence, bioaccumulation, and toxicity, in conjunction with other information about potential exposures and hazards, to help identify chemicals of concern for risk assessment and/or management.

There are significant variations in the way that different government organizations use information about persistence, bioaccumulation, and toxicity characteristics. For example, the Canadian Chemicals Management Plan (CMP) uses the characteristics as a prioritization tool to identify chemicals that require further risk assessment. The United States has a list of priority PBTs related to the dirty dozen but then uses a scoring system for persistence, bioaccumulation, and toxicity (and other characteristics) to prioritize chemicals now in commerce for accelerated exposure and risk assessment of specific uses. In Japan and Europe, the placement of chemicals on official lists of PBTs are for a very different purpose: creation of binding risk management obligations on producers and/or users. In Europe, for example, the mere placement of a potential PBT on the Candidate List of substances of very high concern triggers some reporting obligations in the supply chain, even though substances on the Candidate List are not necessarily placed on the final Authorization List. As a result, when comparing how different countries are conducting PBT determinations, it is useful to have a broader understanding of how those countries are using the PBT determination in the risk assessment and management processes.

Relevant principles: precaution, priority setting, exposure and risk assessment

6. Progress has been made in the international harmonization of the tests and cutoff values used to determine persistence, bioaccumulation, and toxicity, but significant inconsistencies (e.g., the cutoff values and the specific testing requirements) between jurisdictions remain that do not have an identifiable scientific rationale.

There are many tests that can be used to determine persistence, bioaccumulation, and toxicity. In some regulatory contexts (REACH and Toxic

Substances Control Act [TSCA]), Organization for Economic Cooperation and Development (OECD) test guidelines are acknowledged as sound science (especially for persistence and bioaccumulation). More broadly, the OECD test guidelines are often referred to by different organizations as the standard approach to determine persistence, bioaccumulation, and toxicity. However, the tests do vary somewhat, and different regulatory programs use divergent cutoff values. The Stockholm Convention cutoff values capture fewer chemicals than do the values used in REACH. Other programs, such as Japan's, use a persistence *or* bioaccumulation approach (so-called partial PBTs). These variations, which cannot be reduced based on science alone, make it complicated for governments to fully harmonize PBT determinations.

Relevant principles: information quality, value of information, priority setting

7. There are many similarities in how information about persistence, bioaccumulation, and toxicity is used to set priorities for risk assessment and management, but inconsistencies remain not only in the cutoff values and tests used for PBT determinations, but also in the weights assigned to the three properties and related information about potential releases and exposures.

Regulators in several nations use PBT information to set priorities for chemicals that require further exposure and risk assessment. PBTs present challenges for assessment owing to the inability to completely translate modeling and laboratory test studies into real-world environmental situations. In other words, the different environmental contexts that a chemical may enter can influence its persistence and bioaccumulation potential.

Given this complexity, we have found that regulators place different degrees of emphasis on the three properties. US Environmental Protection Agency's (US EPA's) Work Chemicals Plan scoring system combines persistence and bioaccumulation as one score, whereas hazard (toxicity) is scored separately. The US EPA therefore appears to place greater emphasis on toxicity than on persistence or bioaccumulation. REACH, on the other hand, employs a precautionary very persistent, very bioaccumulative (vPvB) criterion as well as a PBT criterion (p. 69). Compared with US regulators, European regulators appear to give more weight to persistence and bioaccumulation, relative to toxicity.

One possible explanation for the difference between the United States and Europe in this regard is that US EPA may use the PBT determination primarily to protect human health while REACH may envision the PBT determination primarily as a tool to protect environmental quality. The difference in emphasis should not be overstated, since both US and European regulators are concerned about protecting both human and environmental health. Indeed, the scoring/priority systems used by US EPA and Canada both consider environmental as well as human health impacts. Nonetheless, the relative weight given to human versus environmental health may not be identical on the two sides of the Atlantic.

Relevant principles: information quality, priority setting, value of information

8. When PBT determinations are used only for screening and priority setting, one might expect that the cutoff values would be more conservative (i.e., inclusive of more chemicals) than when the PBT determinations trigger binding risk management obligations without further assessment. However, no such pattern seems to exist.

Under REACH and Japan's Chemical Substances Control Law (CSCL), PBT determinations trigger relatively strict regulatory consequences. If a chemical is classified as a PBT, then production and distribution are automatically limited, and industry may apply for exceptions for particular uses under narrow criteria. In the United States, formal PBT determinations are not made for existing chemicals, but information on persistence, bioaccumulation, and toxicity is used in a priority-setting scheme to guide further exposure and risk assessments that inform regulatory decisions about whether (and what type of) risk management is needed.

Canada is an intermediate case. Canadian regulators use the PBT determination both as a prioritization tool to guide further screening level risk assessment and, alongside exposure data, as a risk management tool. It might be expected that the screening PBT criteria in Canada might be more inclusive than in Europe (since the European designation has stricter regulatory ramifications), but this is not the case. The persistence and bioaccumulation criteria utilized in Canada are less inclusive (include fewer chemicals) than the REACH criteria. In fact, REACH's vPvB criteria are the same criteria that are used by many other regulatory agencies to identify PBTs, and the REACH regulation requires use of more inclusive cutoff values for PBTs (to include more chemicals) than many other programs.

Relevant principles: information quality, precaution, priority setting, value of information

9. Some jurisdictions go beyond PBT determinations and identify as possible concerns chemicals that might be called "partial PBTs," which typically means that only two of the three properties have been established (e.g., persistence and toxicity or bioaccumulation and toxicity).

Some jurisdictions examine "partial PBTs" in addition to PBTs. In its categorization effort, Environment Canada prioritized PiTs and BiTs for further examination in addition to PBiTs (where iT refers to inherent toxicity). Japan's CSCL applies a limited risk management measure, described above, to PTs (Class II Substances) and PBs (Monitoring Substances) for which toxicity is uncertain. The state of Oregon explicitly chose to include partial PBTs in its Priority Persistent Pollutant List. This is similar to the vPvB designation under REACH. For vPvB substances, however, the toxicity of the chemical is assumed since it is not easy to test such chemicals for toxicity.

Relevant principles: precaution, priority setting

B. Recommendations on PBT determinations

1. To avoid duplication of effort and foster predictability in global trade, greater efforts at international harmonization are needed in data gathering and assessment, and especially in the cutoff values for persistence, bioaccumulation, and toxicity, including the weights given to each characteristic.

To promote predictability, efficiency, and transparency, there is a need for more global harmonization of PBT data gathering and assessment. Cutoff values in different settings were selected primarily through consensus and science-policy judgment rather than empirical science. Harmonizing the cutoff values across organizations would advance the process of PBT determinations, assuming that there is also movement toward harmonization in how the PBT determination is used by regulators.

The testing procedures for persistence and bioaccumulation are a good example of harmonization. In many programs, the OECD test guidelines are used for laboratory testing of chemicals. The OECD uses a consensus-based approach, and therefore, OECD test guidelines are often applied globally. This consensus approach is a good way to create harmonization, and it should be extended to cutoff values and other aspects of PBT determination (e.g., toxicity testing requirements). To move in this direction, OECD will need to take a bit more risk as an organization by moving from strictly scientific work into the science-policy arena and even into regulatory policy harmonization.

Relevant principles: information quality, precaution, exposure and risk assessment, value of information

2. WOE can be a superior approach, but guidance should be developed to help bring more rigor, credibility, and predictability to WOE judgments made in the PBT determination process.

The use of WOE to make PBT determinations needs to be more robustly developed and then explained more clearly across countries. Guidance developed with stakeholder input and feedback would help make this process more transparent and could alleviate some of the unpredictability in PBT determinations. The most important guideline would be to plainly define how WOE, including different lines of evidence, will be derived and then evaluated within different regions or jurisdictions. There are various definitions for WOE, and it is important to understand how this term will be applied and what criteria will guide determinations. We suggest that WOE be applied to persistence, bioaccumulation, and toxicity individually as well as to their relationship to each other. This will allow for all credible data to be considered, thereby promoting more scientifically informed determinations of PBTs.

Relevant principles: information quality, priority setting, value of information

3. Harmonized PBT determinations should be encouraged, including those based on WOE judgments, which implies more regular international dialogue on PBT determinations (with respect to both general procedures and specific substances).

Some PBT determinations are complex because many variables and data sets may be worthy of consideration. Rather than each jurisdiction tackling this complexity on its own, there should be regular international dialogue on both the procedures for making PBT determinations and the WOE determinations for specific substances. As a practical matter, some organizations are likely to insist on distinct procedures or substance-specific opinions, but such differences may be attenuated with regular international dialogue.

Relevant principles: information quality, precaution, priority setting, value of information

4. Biomonitoring programs have clear value in problem identification and in the evaluation of risk management programs, but guidance is needed on how biomonitoring data can be used wisely in making WOE judgments about whether a chemical is a PBT.

As guidance for WOE is developed, information from monitoring and biomonitoring should be addressed in terms of how it is (or is not) relevant to PBT determinations. For example, the mere presence of a chemical in the environment is not enough to consider that chemical persistent or bioaccumulative, particularly if the chemical is repeatedly released into the environment. A guideline that describes how to distinguish a chemical that is persistent from a chemical that has constant releases would increase clarity and accuracy in the use of monitoring data for PBT determinations.

Relevant principles: information quality, precaution, priority setting, exposure and risk assessment, value of information

5. If an initial PBT determination is based on limited or imperfect information, mechanisms are needed to revisit such a determination as new information becomes available. PBT determinations should not be used as a substitute for scientific investigation of the causal mechanism and pathways by which a chemical is persistent, bioaccumulative, or toxic. This implies that the need for causal studies is not diminished once a substance has been listed (or not listed) as a PBT.

Regulators often make potential PBT determinations based upon screening data and Quantitative Structure–Activity Relationships (QSARs). To more accurately identify PBTs, regulations should include a mechanism to review chemicals when new information comes to light. A significant portion of chemicals lack laboratory information or other forms of data.

Siloxane D5 is a good example of how new information can improve and possibly modify (or reverse) an initial PBT determination (see case study in Chapter 3). The mechanism could range from an informal science advisory board to the more formal Board of Review employed in Canada in the case of Siloxane D5. Once a chemical is classified as a PBT, there should be consideration of further testing that clarifies potential risks to human health and the environment. On the other hand, the availability of new information on a non-PBT could also be used to classify it as a PBT.

Relevant principles: information quality, precaution, priority setting, value of information

6. Procedures need to be further developed to ensure periodic review and modernization of the entire PBT determination process based on advances in scientific methods and knowledge (e.g., new metabolism and excretion information, alternative toxicity tests, and improved methods for making WOE judgments).

When designing regulatory programs around the PBT concept, authorities originally thought that there would only be a few dozen such chemicals. Given the increased number of potential PBTs, the process should be modernized in a manner that allows for efficiently and wisely addressing hundreds of chemicals rather than a few dozen. In Europe, the PBT Expert Group is a constructive step in this direction. OECD might consider formation of a similar group that has broader international representation. A well-designed program would also provide a way to prioritize the PBTs that may pose the highest threat to human health and the environment (e.g., because of widespread use, release, and exposure).

Relevant principles: information quality, precaution, priority setting, value of information

7. Given the growing global importance of PBT determinations and exposure/risk assessment processes for risk management around the world, governments and industry need to make larger investments in research programs to generate valid, relevant, and representative data about persistence, bioaccumulation, toxicity, and exposure. These types of investments could advance sustainable molecular design and synthesis of non-PBT substances while also improving the PBT determination and exposure/risk assessment processes.

Many chemicals do not have laboratory or field data to inform PBT determinations. This often results in heavy reliance on QSARs or other estimation methods for making a PBT determination. Even proponents of estimation methods acknowledge that they can be a fallible basis for making such a determination. The process would become more accurate and credible if measurements on these chemicals were generated and made widely available. Moreover, regulatory requirements support the need for this research agenda. The Stockholm Convention, for example, requires that bioaccumulation determinations in risk profiles be based on "measured values" for bioconcentration or bioaccumulation factors.*

Looking forward, we suggest that the major research needs regarding the PBT determination process be identified and targeted funding be directed at solving those needs. This could include not only research on specific chemicals but also development of better methods or analytical tools and processes for retrospectively making PBT determinations and prospectively designing more sustainable chemicals. Moreover, the data required for PBT assessment are also required for exposure and risk assessment. If approximate estimates for chemical release and use can be obtained, then the difference between screening level PBT assessment and screening level risk assessment could be narrowed (Arnot, Mackay, and Bonnell 2008).

Relevant principles: information quality, precaution, priority setting, value of information

8. Chemicals found to be partial PBTs should not be given formal designations because the number of chemicals with all three PBT properties will be high enough for the foreseeable future to stretch the resources of regulators and industry. In addition, regulatory authorities in most jurisdictions studied here have risk management tools

---

* Stockholm Convention, Annex E(c).

available to address the risks posed by partial PBTs (e.g., chemicals that are persistent and toxic and present a high potential for exposure to humans or the environment) without the application of any formal legal designations.

Partial PBTs may be of concern but should not usually be a priority over substances with all three PBT properties. The combination of all three properties presents unique concerns. Therefore, substances with only two of the three properties may not be as worrisome, although policy makers and regulators should also account for other factors (e.g., exposure potential) in prioritization decisions.

Relevant principles: precaution, priority setting

C. Findings on PBT policies

1. The protective goals of health, safety, and environmental policy are broadly shared internationally, but there is wide variation in how policy makers around the world make use of information about persistence, bioaccumulation, and toxicity.

Canada, the European Union (EU), Japan, the United States, and other jurisdictions share commitments to protective policies toward chemicals with PBT properties. Some jurisdictions rely primarily on established international programs (e.g., the Stockholm Convention), while others have initiated their own programs.

There is a wide range of policy responses: modifications to material safety data sheets, voluntary or mandatory risk management measures, mandatory disclosure of releases to the public, incentives for safer chemicals, and prohibitions of PBT manufacture and use. Jurisdictions vary on the policy responses that are emphasized.

Relevant principles: precaution, differentiation, rational alternatives

2. It is well known that the extent of chemical release and exposure is crucial in determining the degree of human health and environmental risk from specific industrial uses, but such information plays a stronger role in some policy schemes than others, and in some jurisdictions, release and exposure information appears to play little or no role in PBT policies.

Risk is described as a combination of exposure, hazard, and dose–response evaluation. In some regulatory organizations (e.g., REACH and Japan's CSCL), exposure is not always considered when making management decisions associated with PBT determinations, in part because the persistence and bioaccumulative properties are used as surrogates for exposure. More recently, the European Commission, under REACH, is beginning to include information about releases during specific uses of chemicals in the Risk Management Options (RMO) process, before a final PBT determination. Canada and the United States include release and exposure information for specific uses when conducting risk assessments for priority chemicals (including chemicals that have been shown to exhibit persistence, potential for bioaccumulation, and toxicity).

While some jurisdictions assume that no safe level of exposure exists for PBTs, this is not a standard assumption in North America. If no safe level of exposure is assumed, then the case for specific uses must rest on a comparison of benefit versus risk, a comparison of risk versus risk of the substitute, or some form of socioeconomic or cost–benefit analysis. Any of these cases will require information about exposure and risk in a use-specific context.

Relevant principles: precaution, exposure and risk assessment, differentiation, value of information

3. Some policies use PBT determinations to help set priorities for risk assessment and management, while other policies use PBT determinations to trigger binding risk management obligations without further assessment.

US EPA uses PBT properties as part of a larger scoring system to help set priorities for chemical risk assessment. Canada also uses PBT information in priority settings. If a chemical is a priority, then a more in-depth assessment is undertaken of specific uses, releases, exposures, risks, management measures, and possible substitutes. Canada also uses the PBT concept, along with exposure data, to inform its risk management decisions. In other programs, such as REACH, the PBT determination can trigger the substitution process, and any use-specific risk assessments for authorization are the obligation of industry. In Europe, PBT assessment is sometimes considered an alternative to full risk assessment, whereas in North America, PBT assessment is a potential precursor to risk assessment.

Relevant principles: precaution, priority setting, exposure and risk assessment, differentiation, rational alternatives

4. Across the jurisdictions studied, there is much duplication of effort in analyses conducted to inform the development of PBT policies (e.g., similar analyses of chemical inventories and similar risk profiles for specific substances).

Each jurisdiction typically conducts its own evaluation or PBT determination on individual chemicals. This is a redundant process that consumes scarce public and private resources. In the case study on Siloxane D5 (see Chapter 3), we found that both Canada and the EU devoted substantial resources (scientific and political) to resolving the PBT determination question for the same chemical. Since governments will be faced with making hundreds of PBT determinations over the next decade, it may not be efficient to have multiple governments working independently on the same chemicals. A more collaborative approach may result in better use of limited government resources while also producing determinations that are more credible with publics on a global basis.

Relevant principles: priority setting, exposure and risk assessment, value of information

5. Comparative risk analysis of chemical alternatives to PBTs, while present in some policies (e.g., parts of the REACH regulation in Europe), is not yet playing a prominent role in risk management of PBTs.

Comparative risk analysis of substitutes for PBTs is envisioned in the REACH authorization process and restrictions procedure. It is not prominent in Japanese policy. In Canada, there is no explicit mention of formal comparative risk analysis between worrisome chemicals and their potential substitutes.

Any rational system of alternatives assessment should be concerned with properties and risks of a possible substitute chemical as well as the properties and risks of an existing chemical on the market. While this principle is widely accepted, its practical application to regulation of industrial chemicals is not straightforward. Under REACH's authorization process, for example, producers of an existing chemical are permitted to sell, on a temporary basis, a substance of very high concern (e.g., PBT) until a suitable substitute becomes available. Does such a producer have adequate economic incentive to undertake research and development on potential substitutes? If not, reliance must be placed on research and development at competitor companies, but their role in the authorization process has not yet been clarified.

Relevant principles: exposure and risk assessment, rational alternatives, value of information

6. No clear-cut framework has been developed to guide determinations as to whether the benefits of PBT use in a particular application justify the potential risks.

The Japanese system is silent on risk–benefit issues. Europe's REACH authorization allows temporary use of PBTs when a socioeconomic analysis shows that the benefits outweigh the risks and there is no suitable substitute for a specific application. This process, which places the analytic burden on industry, is largely untested at this point because it has not been implemented. As REACH's authorization process unfolds on a larger number of chemicals, priority should be given to evaluating how the process works and how it might work better in accomplishing the multiple goals of REACH. Regulatory agencies in the United States and Canada are authorized to perform cost–benefit analyses, but their current application to PBTs is limited. The Stockholm Convention also calls for this information to be supplied before the listing of additional chemicals.

Relevant principles: exposure risk assessment, differentiation, rational alternatives, value of information

7. Although the evidence is limited and does not account for some of the recent regulatory focus on PBTs, it does not appear that the share of PBTs in the industrial inventory is declining over time.

There have been a number of screening studies to identify PBTs among existing and new chemicals. These studies show that there has not been a decrease in PBT use over time. However, the regulatory and industrial focus on PBTs is relatively recent, and thus, it may be too early to detect any decline in PBT use.

Relevant principles: information quality, precaution, priority setting

D.   Recommendations on PBT policies

1. For a variety of scientific and policy reasons, more focus should be given to using PBT determination as a priority-setting tool for risk assessment and management rather than as a trigger for binding risk-management obligations.

Originally, it was thought that there would only be a couple of dozen PBTs among those chemicals now in commerce. With so few in number, it was thought that these chemicals could be phased out or prohibited without much economic consequence. Now that we know that there may be 100 to 1000 PBTs, with literally thousands of distinct uses in the industrial, healthcare, and consumer sectors, it may prove unwise to phase out all of these chemicals, particularly those that lack safer, suitable substitutes. As a result, regulatory policy should move toward a screening approach where persistence, bioaccumulation, and toxicity are used to inform priority setting for risk assessment and management.

Relevant principles: precaution, priority setting, exposure and risk assessment, value of information

2. Policy decisions about PBTs should be informed by and formulated based on the procedures of risk assessment; comparative risk analysis; overall cost–benefit analysis; fairness of the distribution of costs, risks, and benefits; transparency; consultation; peer review; public comment; and formal governmental response to public comment.

PBT policy decisions, while often rooted in the principle of precaution, should be based on high-quality information and a careful consideration of risks, costs, and benefits. When decisions about PBTs are pending, a number of stakeholders may have relevant information and insight about the chemicals under review. With a formal consultation, public comment, and formal government response to public comment, new information and insights on chemicals can be shared with the regulators and policy makers. Under REACH, for example, more public comment procedures should be incorporated into the RMO process as it is refined and formalized. Consultation with all relevant stakeholders as well as the public will also aid regulators and policy makers in accounting for the distributional effects of their decisions.

Relevant principles: information quality, precaution, exposure and risk assessment, differentiation, rational alternatives

3. Policy schemes for PBTs, including risk management programs, should be discerning enough to distinguish, for example, low-benefit applications involving high release and exposure from low-release, high-benefit applications.

When a PBT is used in multiple applications, the risk and benefit of the chemical may vary widely by application. There is an important difference between, for example, uses that have high releases and small benefits versus uses that have low releases and large benefits. Policy schemes should be designed with sufficient flexibility to account for risk–benefit differences by application.

Relevant principles: exposure and risk assessment, differentiation

4. Creative regulatory approaches should be adopted to reward companies that significantly reduce risk, whether through refinement of risk management programs or by proactively replacing PBTs with safer chemicals in the specific uses that have high releases, risks, and exposures.

If industry received some sort of reward for replacing PBTs, then it would be expected that use of—and exposures to—PBTs would decrease over time. PBTs with high potential for release and environmental exposure are chemicals that have the potential to have high risk. Even if it is not feasible to ban these chemicals in the near-term, incentives should be created for industry to replace them with safer substitutes (or enact other risk reduction measures) as soon as possible, especially when the uses have potential for high exposure.

Relevant principles: precaution, differentiation, rational alternatives, value of information

5. Better tools of green and sustainable chemistry are needed to identify chemical innovations that can meet the needs of industry and consumers while minimizing risks to health, safety, and the environment from exposure to PBTs.

When identifying a potential use for a chemical, an analysis should be undertaken to determine if the chemical may potentially be a PBT. Green and sustainable chemistry can help with identifying safer substitutes for risky uses of PBTs before the PBTs make it to the market. If PBTs are marketed, green and sustainable chemistry can also help accelerate the introduction of safer substitutes. US EPA, for example, has developed the PBT Profiler, which allows users to screen chemicals for potential PBT properties.

Relevant principles: rational alternatives, value of information

6. More information on human and environmental exposures to PBTs and actual adverse effects arising from specific uses, including data on the magnitude, frequency, and duration of exposure, is needed to better inform policy decisions about PBTs.

Currently, there is limited information on how a PBT, once it is released, becomes a contributor to human and environmental exposure. If a chemical is detected in human blood, it can be difficult to find the original source and the mechanism by which people were exposed. Since exposure plays a vital role in risk assessment and management, it is important to have a better understanding of how these chemicals are released and transported in the environment. Better data on the dynamics of release, fate, and exposure would provide vital information to policy makers and shed light on promising chemical substitutes. Likewise, data on actual adverse effects caused by PBTs are needed as well as an understanding of the causal mechanisms and pathways. Both industry and government need to make a larger commitment to exposure assessment of specific uses of PBTs.

Relevant principles: information quality, priority setting, rational alternatives, value of information

7. Recognizing that governments may make precautionary decisions when faced with uncertain evidence, international dialogue and

cooperation between governments are advisable before binding risk management decisions are made about specific uses of PBTs. Cooperation ensures that all relevant information has been properly reviewed and the relative costs and benefits of the proposed decision to stakeholders and society are understood.

International dialogue between governments about PBTs is advisable because a consensus decision would be beneficial for clarity, for global protection of the environment and human health, and for international trade. Moreover, the risk management measures adopted to curb releases in one country may also be applicable elsewhere. To some extent, formal international treaties/conventions and international organizations (e.g., the persistent organic pollutant [POP] process under the Stockholm Convention) have fostered a degree of international consensus. But it can take literally decades to negotiate, enact, ratify, and implement formal international agreements. More nimble and timely mechanisms are needed to foster intergovernmental deliberations on PBT issues. A modest step forward would be more informal communication and information sharing between North America, Europe, China, and Japan. More ambitiously, OECD could expand its role as a convener to include regulatory policy as well as science, recognizing that each participant would, in the final analysis, retain its sovereign power.

Relevant principles: information quality, precaution, priority setting, differentiation, rational alternatives

8. New approaches that allow for the sharing of PBT information between governments should be established because they will be helpful in fostering regulatory cooperation, mutual recognition, and harmonization.

Many international agreements, including the Stockholm Convention and the Strategic Approach to International Chemicals Management (SAICM), include information sharing provisions.* In fact, many national and regional programs (e.g., REACH) also include provisions designed to facilitate information sharing; however, those provisions are quite limited in practice. One of the complications here is that test data on a substance may be considered confidential business information, and such data may also have market value (e.g., if one firm has paid for the test data, and a competitor has not, the competitor might seek a way to obtain access to test data without paying for it). When governments consider sharing data on chemicals that have been obtained from companies, consideration needs to be given to whether the data are confidential business information and whether the data have market value. In both cases, governments

---

* Stockholm Convention, Art. 9; SAICM, Overarching Policy Strategy, ¶¶ 15, 17.

need to explore ways to share data while respecting other important interests. In some cases, companies may be willing to permit governments to share their data (or portions of their data) without any compensation.

Relevant principles: information quality, value of information

9. A neutral convener is needed to foster more international dialogue on the development of public policies toward PBTs.

It is natural for governments to prefer making their own policy decisions when it comes to PBTs. But fragmentation of PBT policies is already occurring and may cast doubt on the credibility of national policies (why is a PBT banned in one country and not in another?) and serve as a barrier to international trade and investment. A neutral convener may be needed in some cases to help mitigate some of the policy disagreements.

Relevant principles: information quality, priority setting, differentiation, rational alternatives, value of information

II. Harmonization

We conclude with a modest proposal for regulatory cooperation among Canada, the EU, Japan, China, and the United States. Instead of seeking a formal international treaty in the short-term, we suggest informal government-to-government dialogues on a variety of PBT-related issues. In the long-term, a formal international treaty may be appropriate.

With regard to PBT determinations, the work of OECD and the Stockholm POPs process have led to a fairly workable consensus on technical approaches to identifying potential PBTs. The standard tests and estimation methods for persistence, bioaccumulation, and toxicity, coupled with traditional cutoff values, should facilitate PBT determinations for the vast majority of chemicals now in commerce (see Rauert et al. 2014). If effective guidance is established, it is appropriate to employ WOE considerations in the minority of cases where there are unusual patterns of data or seemingly conflicting information.

Before PBT determinations are made on specific substances, we recommend that a regulatory cooperation consultation occur among the world's key regulatory scientists. In the near-term, Canada, the EU, Japan, and the United States could lead the effort. The purpose of the cooperation consultation is to avoid, whenever possible, conflicting PBT determinations among the four jurisdictions. We recommend that the consultation occur informally, within the current legislative and administrative frameworks that are now in place in these jurisdictions. If these four leaders in the production, use, and regulation of chemicals can harmonize their decisions, it would create a model that other industrialized and developing countries could, and likely would, follow.

Achieving harmonization of PBT policies may be more elusive, but we believe that more effort at regulatory cooperation within current legislative frameworks is worthwhile. Canada, the EU, Japan, and the United States are moving at somewhat different paces on PBT policy, but all four jurisdictions seem likely to give greater priority to PBTs over the next decade or two. Disparate regulatory outcomes are likely to create public confusion, further increase investment uncertainty, and result in unnecessary costs for companies operating in a global market. Thus, the four jurisdictions are encouraged to engage in a regulatory cooperation consultation before allowing disparate regulations of a PBT to be implemented. Some regulatory differences may be unavoidable, but more efforts at consultation seem appropriate than are now taking place. In fact, we found remarkably little evidence that such intergovernment dialogues are underway on a regular basis.

Since the EU's REACH and Japan's CSCL programs are far more prescriptive than the Canadian CMP and US TSCA programs, one might think that any effort at regulatory cooperation is destined for failure. We believe otherwise. When a chemical is considered for placement on the Candidate List under REACH on the basis of a PBT determination, the EU has some flexibility (e.g., during the proposed RMO process) to determine the appropriate regulatory response (e.g., restrictions, authorization, or reliance on other regulatory programs). The Japanese system has also evolved to be somewhat more flexible and responsive to international developments.

For most PBTs, it is hard to envision compelling reasons for different regulatory outcomes in the four jurisdictions. Since most PBTs will have multiple industrial uses that are common to Canada, the EU, Japan, and the United States, we recommend that these uses be a subject of regulatory dialogue and cooperation, as all four systems have discretion in how specific uses of PBTs are handled. The purpose of the consultation is to share information on what is known about the risks, costs, and benefits of the PBTs in specific uses, compared with the risks, costs, and benefits of possible substitutes. During this consultation process, industry and NGO stakeholders should be informed of the cooperative activity and be provided an opportunity to supply relevant information on specific uses. If industry and governments overcome legal obstacles and agree to make some REACH registration information available to officials in all four jurisdictions, the dossiers should provide a solid starting point for cooperative dialogue among regulatory officials (European Commission 2012; ACC 2013; GAO 2013, 2014; Müller 2014; Rauert et al. 2014).

III.   *Modernization*

In the course of our study, some lines of inquiry were identified that, if pursued, could lead to new approaches or conceptual advances in the science and policies of PBTs. Some of these topics have already been the focus of limited research, while others are nascent ideas. We briefly describe these topics in hopes of stimulating research that improves the governance of PBTs. What follows is not an exhaustive list of research needs related to PBTs, and they are presented in no particular order.

A.   Intake fraction

The characteristics of persistence, bioaccumulation, and toxicity are not the sole determinants of whether or not a chemical should be subject to regulation. It has been suggested that the addition of "intake fraction" to the PBT concept would strengthen the determinations by directly considering exposure. Intake fraction is the proportion of a released chemical that ultimately reaches people or other species via inhalation, ingestion, or dermal exposure (Bennett et al. 2002b). Intake fractions for a given chemical can differ substantially in different uses. For example, a substance released indoors in an urban setting is likely to have much greater exposure than if it is released outdoors in a rural environment. The role that intake fraction might play in chemical risk assessment has been examined (Bennett et al. 2002a; Li and Hao 2003; Margni et al. 2004), but explicitly integrating the intake fraction into PBT determinations represents a promising new line of applications research.

B.   Characterization and assessment of UVCBs
and multiconstituent substances

Assessment of substances with "unknown or variable composition, complex reaction products or biological material(s)" (UVCBs) is extremely challenging in the context of PBT determinations. Multiconstituent substances (MCSs) also present challenges but often not to the same extent as UVCBs. Regarding persistence, the European Chemicals Agency (ECHA 2014) states that "one cannot easily assess the persistence of complex substances that contain many constituents using biodegradation testing methods that measure summary parameters (e.g., CO_2 evolution), since these tests measure the properties of the whole substance but do not provide information on the individual constituents" (p. 78). Similar concerns exist for bioaccumulation: "most bioaccumulation test methods are not applicable (or at least difficult to apply) to MCS or UVCB substances" (ECHA 2014, p. 79). Canadian authorities assume that all UVCBs are persistent and therefore do not directly test for persistence (or bioaccumulation) and proceed to toxicity testing (Ginsburg et al. 2005). There is a clear need for test methods for persistence and bioaccumulation that are applicable to UVCBs and MCSs. Research targeted at filling this need could significantly strengthen the scientific basis for PBT determinations involving UVCBs and MCSs.

C.   Development of an improved numeric scoring system for PBTs

The PBT determination process produces a yes/no outcome for each chemical under consideration, but it does not rank PBTs. Additionally, in most determination schemes, persistence, bioaccumulation, and toxicity each carry equal weight in the process (an exception being the vPvB category used under REACH, where toxicity is not given any weight and extreme values of persistence and bioaccumulation are given special weight).

For purposes of prioritization, there would be some value in an improved scoring system that would enable the relative ranking of PBTs. The United States' Waste Minimization Priority Tool and TSCA scoring tool, Japan's Priority Assessment Chemical Substance system, and the Scoring and Ranking Assessment Model may be useful starting points upon which to base improvements. There might also be situations in which one or two of the three properties should be given greater weight than the other(s).

Toward this end, it was suggested that there would be value in a continuous-variable index of persistence, bioaccumulation, and toxicity that reflects the relative contribution of each property to overall risk potential. In other words, the score is intended as an overall, integrated proxy for risk. Beyond ranking chemicals, this type of scoring system could be used to rank specific uses of chemicals by making use of differences in release rates and exposures.

US EPA currently uses an additive scoring system under TSCA and differentially weights persistence, bioaccumulation, and toxicity (see Chapter 5). Despite US EPA's activities on this front, it is unknown if an additive or multiplicative approach is better, and there are many potential approaches to differentially weighting the three characteristics. At present, the selection of approaches appears arbitrary, and considerable research is needed in this area if a scientifically defensible, numeric index is to be developed.

D.   Development of WOE training programs
for regulatory professionals

As discussed in Chapter 3, use of WOE in PBT determinations is increasing. We have argued for the development of guidelines on how to rigorously and consistently apply a WOE approach in PBT determinations. In addition to technical guidance, there could be value in programs that inform and train regulatory professionals in how WOE is used in the PBT determination process, how WOE is used in different jurisdictions, and how WOE might be applied in constructing new PBT policies. Development of these types of training programs would require investment of time and resources, but doing so could make a meaningful contribution to global harmonization of PBT determinations and policies.

E.   *Uncertainty in WOE*

All measurements have some degree of uncertainty, and the manner in which uncertainty is addressed affects the confidence in determinations and policies based on those measurements. In PBT determinations, it appears there is no harmonized approach to addressing uncertainty. In fact, uncertainty is often given no formal treatment at all.

A WOE approach is, in essence, an attempt to account for uncertainty. But WOE does not quantify the uncertainty in an empirical or probabilistic manner, nor does it address the possibility that uncertainty in the individual tests for persistence, bioaccumulation, and toxicity could be compounded in the final PBT determination. Development of a method for formal (i.e., statistical) treatment of uncertainty specifically for PBT determinations would allow regulatory professionals to assign an appropriate degree of confidence to PBT determinations. Formal treatment of uncertainty also reveals which data are most variable and which types of tests tend to produce variable data, thus identifying those testing methodologies that should be improved.

IV.   *Conclusion*

The diffusion of policy innovations across borders has been well documented in a variety of contexts, especially for environmental and public health regulation (e.g., Vogel 1995; Vogel and Swinnen 2011). In the United States, it is relatively common for state governments to replicate or refine policies that were originally adopted in other states. Likewise, the federal government often borrows from the experiences of one or more of the 50 states.

The history of the PBT concept is itself a case study in policy diffusion. Originally adopted in Japan in 1973, the concept has since been employed in other countries and by international organizations. They have used persistence, bioaccumulation, and toxicity as prioritization factors for risk assessment or to define targets for informational requirements or risk management.

A growing body of literature that documents policy diffusion in the area of chemical regulation also draws lessons from comparative examinations (e.g., Applegate 2008; Sachs 2009; Scott 2009; Abelkop and Graham 2015). REACH and the Canadian CMP, in particular, seem to have influenced recent policy developments, to varying degrees, in other jurisdictions such as Japan, China, and South Korea (Naiki 2010; Uyesato et al. 2013). The empirical literature shows that what policy makers do in some parts of the world influences policy makers do in other jurisdictions. To some extent, the sharing may reflect direct communication between policy makers, but there is also communication between stakeholders and academics around the world. Since stakeholders and academics also influence policy development, the mechanisms by which policy diffusion occurs are complex. Overall, we see policy diffusion in the governance of industrial chemicals as a positive development, as a process of learning is taking place for products that are used in a global marketplace and for which pollution can cross borders.

PBT policy is advancing at multiple levels of governance. At a global level, policy action is largely driven by the goals expressed in the World Summit on Sustainable Development and the SAICM to achieve the sound management of chemicals by 2020. The Stockholm Convention process for POP identification and regulation is an illustration of global action even though some countries are not yet fully participating. At the regional level, important policy actions have taken place in the Great Lakes region and in Europe. At the national level, pioneers have included Japan and Canada, with the United States and China seemingly poised to consider how to proceed in the future. Subnational governmental jurisdictions, from Ontario in Canada to California in the United States, have also emerged as important players in PBT policy activity.

The proliferation of lists of PBTs around the world has both positive and negative ramifications. If criteria are transparent and based on evidence-based guidelines, it is feasible for jurisdictions to learn from each other in the PBT identification process and work, whenever feasible, toward harmonized decisions. On the other hand, when PBT lists are used for different purposes, it is important for policy makers and stakeholders to appreciate those differences and use the information from lists accordingly. When PBT lists are not clear with regard to their bases and uses, they can become a source of policy confusion, stakeholder conflict, and tension in international trade.

For better or worse, multiple lists of PBTs are here to stay, and they—along with governmental and commercial screening tools—are exerting influence on risk management activities through the chain of commerce. Product manufacturers and retailers are utilizing such information and responding to increased consumer awareness of the chemicals within the products they purchase.

Our investigation has established that, while the PBT concept is subject to uneven degrees of use in the public and private sectors, it is widely used, and we expect its use to expand in the future. Whether through "retail regulation" or national regulatory programs, targeting PBTs is seen as a way to positively advance green chemistry by discouraging the development and use of chemicals with persistence, bioaccumulative, and toxic properties. Because large data gaps exist for the tens of thousands of chemicals already in commerce, and limitations exist on public resources to assess them, it is not feasible for governments to fully characterize the risks associated with all of these chemicals within the foreseeable future. Thus, in addressing risks from chemicals already in commerce, the PBT concept is most widely—and we think most wisely—used to identify priority chemicals for risk assessment and potential management.

As we have noted, diffusion does not necessarily entail harmonization. Inconsistencies in PBT determinations and policies can be problematic if they trigger premature retail regulation that does not adequately consider the risks of substitutes, or if they generate unnecessary international trade disputes (e.g., Kogan 2013). One noteworthy development is the potential recognition by one government of another government's chemical classification—a feature of California's Safer Consumer Products Regulations. This practice conserves public resources by minimizing unnecessary duplication of PBT assessments across jurisdictions. However, differences in the way that different jurisdictions define persistence, bioaccumulation, and toxicity could undermine the credibility of this approach. The credibility of this practice should increase as the science that underlies PBT determinations (and other chemical classifications) advances, as WOE guidelines for PBTs are formulated and applied to multiple substances, and as information sharing becomes more commonplace between governments, industry, and NGOs.

Because PBTs and their close cousin, the partial PBT, remain somewhat fluid regulatory constructs, opportunity exists for a greater degree of cross-country and international collaboration on PBT identification and policy. If policy diffusion occurs wisely, the ultimate result may be health and environmental progress around the world as well as a reduction in unnecessary barriers to international trade.

References

Abelkop, A.D.K., and J.D. Graham. 2015. "Regulation of Chemical Risks: Lessons for Reform of the Toxic Substances Control Act from Canada and the European Union." *Pace Environmental Law Review* 32.

ACC (American Chemistry Council). 2013. "ACC Responds to GAO's Report on EPA's Regulation of Chemicals under TSCA." *ACC News Releases*, May 3. Available at http://www.americanchemistry.com/Media/PressReleases Transcripts/ACC-news-releases/ACC-Responds-to-GAOs-Report -on-EPAs-Regulation-of-Chemicals-Under-TSCA.html.

Applegate, J.S. 2008. "Synthesizing TSCA and REACH: Practical Principles for Chemical Regulation Reform." *Ecology Law Quarterly* 35:721–69.

Arnot, J.A., D. Mackay, and M. Bonnell. 2008. "Estimating Metabolic Rates in Fish from Laboratory Data." *Environmental Toxicology and Chemistry* 27:341–51.

Bennett, D.H., M.D. Margni, T.E. McKone, and O. Jolliet. 2002a. "Intake Fraction for Multimedia Pollutants: A Tool for Life Cycle Analysis and Comparative Risk Assessment." *Risk Analysis* 22:905–18.

Bennett, D.H., T.E. McKone, J.S. Evans, W.W. Nazaroff, M.D. Margni, O. Jolliet, and K.R. Smith. 2002b. "Defining Intake Fraction." *Environmental Science & Technology* 36:207–16.

ECHA (European Chemicals Agency). 2014. *Guidance on Information Requirements and Chemical Safety Assessment, Chapter R.11: PBT/vPvB Assessment, Version 2.0*. Available at http://echa.europa.eu/documents/10162/13632/information_requirements_r11_en.pdf.

European Commission. 2012. "Regulatory Frameworks for Chemicals Need More Harmonising." Science for Environment Policy, DG Environment News Alert Service, Issue 290, June 29. Available at http://ec.europa.eu/environment/integration/research/newsalert/pdf/290na4_en.pdf.

GAO (Government Accountability Office). 2013. *Toxic Substances: EPA Has Increased Efforts to Assess and Control Chemicals but Could Strengthen Its Approach*. GAO-13-249. Washington, DC, March 22. Available at http://www.gao.gov/products/GAO-13-249.

GAO (Government Accountability Office). 2014. *Agencies Coordinate Activities, but Additional Action Could Enhance Efforts*. GAO-14-763. Washington, DC, September 29. Available at http://www.gao.gov/products/GAO-14-763.

Ginsburg, J., K. Khatter, F. de Leon, and S. Sang. 2005. *ENGOs' Comments on the Categorization Process and Environment Canada's Proposal on Polymers and UVCBs*. Toronto, ON, Canada: Canadian Environmental Law Association. Available at http://s.cela.ca/files/uploads/513Polymers_UVCBs.pdf.

Kogan, L.A. 2013. "REACH Revisited: A Framework for Evaluating Whether a Non-Tariff Measure Has Matured into an Actionable Non-Tariff Barrier to Trade." *American University International Law Review* 28:489–668.

Li, J., and J. Hao. 2003. "Application of Intake Fraction to Population Exposure Estimates in Hunan Province of China." *Journal of Environmental Science and Health, Part A* 38:1041–54.

Margni, M., D.W. Pennington, C. Amman, and O. Jolliet. 2004. "Evaluating Multimedia/Multipathway Model Intake Fraction Estimates Using POP Emission and Monitoring Data." *Environmental Pollution* 128:263–77.

Müller, T. 2014. "Screening Criteria for PBTs/vPvBs." *Chemical Watch*. Global Business Briefing, June.

Naiki, Y. 2010. "Assessing Policy Reach: Japan's Chemical Policy Reform in Response to the EU's REACH Regulation." *Journal of Environmental Law* 22:171–95.

Rauert, C., A. Friesen, G. Hermann, U. Jöhncke, A. Kehrer, M. Neumann, I. Prutz et al. 2014. "Proposal for a Harmonised PBT Identification across Different Regulatory Frameworks." *Environmental Sciences Europe* 26:9.

Sachs, N.M. 2009. "Jumping the Pond: Transnational Law and the Future of Chemical Regulation." *Vanderbilt Law Review* 62:1817–69.

Scott, J. 2009. "From Brussels with Love: The Transatlantic Travels of European Law and the Chemistry of Regulatory Attraction." *American Journal of Comparative Law* 57:897–942.

Uyesato, D., M. Weiss, J. Stepanyan, D. Park, K. Yuki, T. Ferris, and L. Bergkamp. 2013. "REACH's Impact in the Rest of the World." In *The European Union REACH Regulation for Chemicals: Law and Practice*, edited by L. Bergkamp, 334–70. New York: Oxford University Press.

Vogel, D. 1995. *Trading Up: Consumer and Environmental Regulation in a Global Economy*. Cambridge, MA: Harvard University Press.

Vogel, D., and J.F.M. Swinnen. 2011. *Transatlantic Regulatory Cooperation: The Shifting Roles of the EU, U.S., and California*. Northampton, MA: Edward Elgar.

Appendix

Names and affiliations of interviewees who provided information and expert opinions regarding PBTs. Also indicated are those who provide comments or peer review of the technical report on which much of this book is based. The people listed do not necessarily agree with the methods, findings, or recommendations in this book.

Kevin Armbrust[†]	Louisiana State University
Jon Arnot[*‡]	University of Toronto
Charlie Auer[*†]	Charles Auer & Associates, LLC
Sylvain Bintein[*]	DG-Environment, European Commission
Vito Buonsante[*]	ClientEarth
Bill Carroll[*]	Occidental Chemical Corporation
Holly Davies[*]	Department of Ecology, Washington State
Dennis Devlin[‡]	Exxon Mobile Corporation
Bob Diderich[*]	Organization for Economic Cooperation and Development
Joop de Knecht[*]	Organization for Economic Cooperation and Development
Peter Dohmen[*]	BASF Corporation
Steve Dungey[*‡]	Environment Agency, United Kingdom
Cathy Fehrenbacher[*]	US Environmental Protection Agency
Christina Franz[*]	American Chemistry Council
Vincenza Galatone[*]	Environment Canada
Mike Gallagher[*]	Department of Ecology, Washington State
John Giesy[*]	University of Saskatchewan
Geoff Granville[†]	GCGranville Consulting Corporation

Mark Greenwood*†	Greenwood Environmental Counsel, PLCC
Joshua Grice*	Department of Ecology, Washington State
Joseph H. Guth*	University of California, Berkeley
Dale Hattis†	Clark University
Ronald Hites*‡	Indiana University, Bloomington
Phil Howard*	Syracuse Research Corporation
David Kent*	Keller & Heckman, LLP
Amardeep Khosla*	Industry Coordinating Group for Canadian Environmental Protection Act
Masaru Kitano*‡	Meiji University
Akos Kokai*	University of California, Berkeley
Eeva Leinala*‡	Health Canada
Peter Lepper*	European Chemicals Agency
Laurence Libelo*	US Environmental Protection Agency
Anna Liisa-Sudquist*	European Chemicals Agency
Jeff Lincer*	Researchers Implementing Conservation Action
Gordon Lloyd*‡	Chemistry Industry Association of Canada
Don Mackay*‡	Trent University
Petteri Mäkelä*	European Chemicals Agency
Kevin Masterson*	Department of Environmental Quality, Oregon
Shigeki Masunaga†	Yokohama National University
Bette Meek*	University of Ottawa
John Moffett*‡	Environment Canada
David Morin*	Environment Canada
Tom Parkerton‡	Exxon Mobil Corporation
Johanna Peltola-Thies*	European Chemicals Agency
Jake Sanderson*	Environment Canada
Linda Santry*‡	NOVA Chemical Corporation
Martin Scheringer†	ETH Zürich
Jennifer Seed*	US Environmental Protection Agency
David Shortt*‡	Dow Chemical Canada ULC

Dick Sijm*‡	National Institute for Public Health and the Environment (RIVM), The Netherlands
Keith Solomon†	University of Guelph
Georg Streck*‡	DG-Enterprise, European Commission
Jose Tarazona*	European Chemicals Agency
Eisaku Toda*	Ministry of the Environment, Japan
Henrik Tyle*	Environmental Protection Agency, Denmark
Rob Visser*‡	Organization for Economic Cooperation and Development (retired)
Mike Walls*	American Chemistry Council
David Widawsky*	US Environmental Protection Agency
Dolf van Wijk*	European Chemical Industry Council (Cefic)
Graham Willmott*‡	DG-Enterprise, European Commission

*Interviewee

†Compensated peer reviewer

‡Commenter

Index

Page numbers followed by f and t indicate figures and tables, respectively.